Progress in Turbulence - Fundamentals and Applications

Volume 1

Series Editors

Joachim Peinke, Institute of Physics and ForWind, Carl von Ossietzky University of Oldenburg, Oldenburg, Germany

Christos Vassilicos, Department of Aeronautics, South Kensington Campus, Imperial College London, London, UK

Martin Oberlack, Department of Mechanical Engineering, TU Darmstadt, Darmstadt, Germany

The series Progress in Turbulence publishes new developments and advances in the Turbulence, rapidly and informally but with a high quality. Turbulence presents a large number of aspects and problems, which are still unsolved and which challenge research communities in engineering as well as in physical and mathematical sciences both in fundamental and applied research. The very wide interest in turbulence is not only based on it being a difficult scientific problem but also on the important influence that turbulent flows have in the technical world and our daily life. The intent is to cover fundamental research as well as research related to applications of turbulence, technical and methodological issues, and multidisciplinary aspects. The series comprises three strands: monographs, conference/workshop proceedings and reviews on progress in turbulence research to appear once every 4 years. Of particular value to both the contributors and the readership are the short publication time frame and the worldwide distribution, which enable both wide and rapid dissemination of research output.

More information about this series at http://www.springer.com/series/16269

Jan Friedrich

Non-perturbative Methods in Statistical Descriptions of Turbulence

Springer

Jan Friedrich
Département de Physique
École Normale Supérieure de Lyon
Lyon, Rhône, France

ISSN 2661-8168   ISSN 2661-8176 (electronic)
Progress in Turbulence - Fundamentals and Applications
ISBN 978-3-030-51976-6   ISBN 978-3-030-51977-3   (eBook)
https://doi.org/10.1007/978-3-030-51977-3

This Springer imprint is published by the registered company Springer Nature Switzerland AG
The registered company address is: Gewerbestrasse 11, 6330 Cham, Switzerland

As she then walked away she did not look back at the hut, but let her gaze drift upwards, let it wander beyond the rag-tag disorder of the settlement into the dark forest. Looked at the night-time blackness above. Looked at the snow falling into the cones of yellow electric light. Watched the snow circling its way earthwards. The way the white flakes eddied and whirled in the air as if they were time passing not constantly but erratically. Maria Buloh watched the way the falling snow showed the air was never still, but held endless circling complexities, held infinite possibilities for graceful inexplicable movements.

Richard Flanagan, The sound of one hand clapping

Preface

The past decades of turbulence research have seen considerable advances in our understanding of the problem of fluid turbulence: Ever-increasing computational power led the way to direct numerical simulations of the Navier-Stokes equation at very high Reynolds numbers. At the same time, experiments which operate in large wind tunnels or with fluids at low viscosities such as normal helium further enlarged the availability of turbulence data. Moreover, these empirical findings have been quantified within modeling approaches by stochastic methods and further phenomenological descriptions of turbulent fluctuations. Nonetheless, one of the central problems in statistical descriptions of turbulence, the so-called *closure problem of turbulence* manifesting itself in form of hierarchical multi-point statistics, has proven to be reluctant to current perturbative methods. This is especially sobering given the fact that such methods were extremely successful in other branches of physics, e.g., renormalization methods in quantum electrodynamics or renormalization group methods in statistical physics. It is therefore at the utmost importance to develop concepts and methods that would contribute to a *non-perturbative treatment* of the nonlinearity and nonlocality in the Navier-Stokes equation.

The present monograph intends to give a compact overview of statistical descriptions of hydrodynamic turbulence and is mainly addressed to physicists and engineers who wish to get acquainted with the prevalent concepts and methods in this research area. First, several perturbative treatments such as the quasi-normal approximation, the renormalization method, and the renormalization group are summarized and their shortcomings and limitations within the field of fluid turbulence are pointed out. Subsequently, we discuss three different examples for non-perturbative approaches: the so-called instanton formalism, the operator product expansion, and a stochastic interpretation of the energy transfer within a turbulent fluid. In order to categorize the various methods within their original fields of use, the monograph possesses special sections designated with an asterisk. These sections are not mandatory and can be omitted during a first read.

As most of the topics covered here emanated directly from my Ph.D. thesis, this monograph is by no means an alternative to standard text books on the problem of hydrodynamic turbulence. These include, without claiming to be exhaustive, the two comprehensive volumes on statistical fluid mechanics by Monin and Yaglom [1], the books on perturbative methods by McComb [2] and Lésieur [3], the introductions for engineers by Pope [4] or Davidson [5], as well as the standard book by Frisch [6]. For a treatment via dynamical systems theory, especially in the context of two-dimensional turbulence, I also refer the reader to the turbulence chapter in the book "An Exploration of Dynamical Systems and Chaos" by Argyris, Faust, Haase, and R. Friedrich [7].

I especially want to thank my Ph.D. advisor Rainer Grauer and Joachim Peinke for their encouragement and support throughout my Ph.D. thesis. Over the course of the past few years I have benefited from many interesting discussions with Alain Pumir, Andre Fuchs, Anton Daitche, Armin Fuchs, Aurore Naso, Bernd Lehle, Fabian Godeferd, Georgios Margazoglou, Holger Homann, Johannes Lülff, Jörg Schumacher, Jürgen Dreher, Laurent Chévillard, Lisa Borland, Luca Biferale, Maria Haase, Martin Oberlack, Michael Wilczek, Mickaël Bourgoin, Nico Reinke, Oliver Kamps, Raffaele Marino, Romain Volk, Thomas Trost, Tobias Grafke, Stephan Eule, and many others. I am grateful to Jürgen Möllenhof for all his technical support. Moreover, I want to thank Holger Homann for letting me use his data from numerical simulations in Chap. 5. I am also thankful to P. K. Yeung who kindly agreed on letting me reproduce Figs. 2.4 and 2.5 from one of his papers. Furthermore, I am indebted to Janka Lengyel for her help with the manuscript.

I want to acknowledge funding from the Alexander von Humboldt Foundation within a Feodor-Lynen scholarship. I also benefited from financial support of the Project IDEXLYON of the University of Lyon in the framework of the French program "Programme Investissements d'Avenir" (ANR-16-IDEX-0005).

Finally, I wish to thank the members of the staff at Springer-Verlag for their professional cooperation, in particular, Thomas Ditzinger and Daniel J. Glarance for their patience and continuous encouragement and support.

Vienne, France   Jan Friedrich
April 2020

References

1. Monin, A.S., Yaglom, A.M.: Statistical fluid mechanics: Mechanics of turbulence. Courier Dover Publications (2007)
2. McComb, W.D.: The physics of fluid turbulence. Oxford University Press (1990)
3. Lesieur, M.: Turbulence in fluids (2012)
4. Pope, S.B.: Turbulent flows. Cambridge University Press (2000)
5. Davidson, P.A.: Turbulence: an introduction for scientists and engineers. Oxford University Press (2015)
6. Frisch, U.: Turbulence. Cambridge University Press (1995)
7. Faust, G., Argyris, J., Haase, M., and Friedrich, R.: An Exploration of dynamical systems and chaos. Springer (2015)

Contents

Chapter 1
Introduction

The physics of turbulence is a fascinating but at the same time inherently complex branch of classical physics. Although the underlying equation, i.e., the Navier-Stokes equation are known for nearly two centuries, we have yet to identify probabilistic methods that would allow for an accurate description of the spatio-temporal complexity exhibited by its velocity field fluctuations [1].

From an academic point of view, the Navier-Stokes equation constitutes the paradigm of a non-equilibrium physical system with large numbers of strongly interacting degrees of freedom. Accordingly, concepts and methods that are used in turbulence theory have a strong influence on other research fields concerned with strongly interacting systems, e.g., quantum chromodynamics.

Nevertheless, the problem of hydrodynamic turbulence is not merely an academic one. In the course of the so-called "Energiewende" in Germany, referring to the transition toward sustainable energy sources, considerable efforts are being made in pursuance of decreasing CO_2 emissions by the year 2050. Achieving these aims, which also include the withdrawal from nuclear energy by the year 2022, heavily relies on the deployment of renewable energies such as wind or solar power. In contrast to the rather steady and controllable power output of a nuclear or combustion plant, wind farms exhibit heavily fluctuating and uncontrollable power output [2]. Latter can be attributed to the turbulent medium in which wind turbines operate. Consequently, a better understanding of probability laws that govern turbulent fluid motion would signify a tremendous step toward better stability of energy grids, improved maintenance of wind turbines, and ultimately toward meeting the total energy demands from renewable energies. Furthermore, it might also affect other areas of public interest, for instance, regarding better weather forecast capacities or general engineering problems, e.g., the construction of aircrafts.

© Springer Nature Switzerland AG 2021
J. Friedrich, *Non-perturbative Methods in Statistical Descriptions of Turbulence*,
Progress in Turbulence - Fundamentals and Applications 1,
https://doi.org/10.1007/978-3-030-51977-3_1

The key problem of turbulence is that the Navier-Stokes equation proves to be reluctant with regard to a perturbative treatment of its nonlinearity. Therefore, perturbative methods that were successful in other branches of physics, e.g., renormalization methods in quantum electrodynamics or the renormalization group from critical phenomena, lead to imprecise or even amiss results. The following chapters are devoted to a statistical description of turbulence in the Eulerian frame, i.e., in a fixed frame of reference, in contrast to the Lagrangian description, which considers a frame of reference co-moving with the fluid. The resulting statistical equations reveal a *hierarchical ordering*: statistical quantities of a given order depend on statistical quantities of higher order, which is referred to as the *closure problem of turbulence*. Common methods of circumventing the closure problem depart from the assumption of Gaussianity of the velocity field. However, an inherent feature of turbulence is that velocity field fluctuations exhibit statistics that strongly deviate from Gaussian distributions. The latter *phenomenon of intermittency* is recognized as the influence of the intense nonlinear and nonlocal character of the Navier-Stokes equation. Therefore, new methods that would allow for a non-perturbative treatment of the nonlinearity in the Navier-Stokes equation have to be developed. In the following, we want to address the difficulties that the Navier-Stokes equation imposes on the realm of a statistical description of turbulence. Therefore, the chapters are organized as follows: Chapter 2 specifies the basic equations that govern turbulent fluid motion. Chapter 3 focuses on a statistical description of turbulence and phenomenological models of turbulence. Evolution equations for statistical quantities, such as correlation functions and multi-point probability density functions (PDF), are derived from the Navier-Stokes. Chapter 4 discusses closure methods for the moment hierarchy such as the quasi-normal approximation, the eddy damped quasi-normal approximation, further perturbative approaches (direct interaction approximation, Wyld diagrams), and methods from field theory (renormalization methods and the renormalization group). Each closure is carefully analyzed and its shortcomings and failures are pointed out. Finally, Chapter 5 discusses three different non-perturbative treatments that were developed within the past three decades: the instanton formalism, the operator product expansion, and a stochastic interpretation of the turbulent energy cascade. The final chapter tries to give an outlook on how these seemingly different methods might further improve our understanding of the problem of turbulence.

References

1. Frisch, U.: Turbulence. Cambridge University Press (1995)
2. Peinke, J., Heinemann, D., Kühn, M.: Windenergie - eine turbulente Sache. Phys. J. **13**(7), 36 (2014)

Chapter 2
Basic Properties of Hydrodynamic Turbulence

This introductory chapter will give an overview of basic hydrodynamic equations and the difficulties they present in the realm of fully developed turbulence. Furthermore, important quantities such as the Reynolds number or the local energy dissipation rate will be introduced.

2.1 The Basic Fluid Dynamical Equations

The temporal evolution of the velocity field $\mathbf{u}(\mathbf{x}, t)$ in a three-dimensional viscous fluid is governed by the Navier-Stokes equation

$$\frac{\partial}{\partial t}\mathbf{u}(\mathbf{x}, t) + \mathbf{u}(\mathbf{x}, t) \cdot \nabla \mathbf{u}(\mathbf{x}, t) = -\frac{1}{\rho}\nabla p(\mathbf{x}, t) + \nu \Delta \mathbf{u}(\mathbf{x}, t) , \qquad (2.1.1)$$

where $p(\mathbf{x}, t)$ denotes the dynamic pressure field and ν the kinematic viscosity. Moreover, the density ρ is assumed to be constant and put to unity for later convenience. Latter restraint imposes a supplementary condition on the velocity field, i.e., the incompressibility condition

$$\nabla \cdot \mathbf{u}(\mathbf{x}, t) = 0 , \qquad (2.1.2)$$

which is a direct consequence of the continuity equation. When supplemented with additional boundary conditions for the velocity field, the nonlinear partial differential equations (2.1.1) and (2.1.2) completely define the mathematical problem of hydrodynamic turbulence. A remarkable property of Eq. (2.1.1) is that it solely involves

© Springer Nature Switzerland AG 2021
J. Friedrich, *Non-perturbative Methods in Statistical Descriptions of Turbulence*,
Progress in Turbulence - Fundamentals and Applications 1,
https://doi.org/10.1007/978-3-030-51977-3_2

the kinematic viscosity v as a molecular property of the fluid. This property can be expressed in a non-dimensional form of the Navier-Stokes equation (2.1.1) in terms of non-dimensional quantities

$$\tilde{\mathbf{u}} = \frac{\mathbf{u}}{U} \,, \quad \tilde{t} = \frac{t}{T} \,, \quad \tilde{\mathbf{x}} = \frac{\mathbf{x}}{L} \,, \quad \tilde{p} = p\frac{L}{U^2} \,, \quad U = \frac{L}{T} \,, \tag{2.1.3}$$

where L is a typical length scale (e.g., the mean diameter of an obstacle immersed in the fluid) and U is a typical velocity (e.g., the mean velocity) in the flow. Subsequently dropping the tilde signs yields the non-dimensional Navier-Stokes equation

$$\frac{\partial}{\partial t}\mathbf{u}(\mathbf{x}, t) + \mathbf{u}(\mathbf{x}, t) \cdot \nabla \mathbf{u}(\mathbf{x}, t) = -\nabla p(\mathbf{x}, t) + \frac{1}{\mathrm{Re}}\Delta \mathbf{u}(\mathbf{x}, t) \,. \tag{2.1.4}$$

Here, the only externally prescribed parameter enters through the non-dimensional Reynolds number

$$\mathrm{Re} = \frac{UL}{v} \,. \tag{2.1.5}$$

Hence, in the absence of external forces, flows with identical Reynolds numbers will be *geometrically similar*. It is therefore possible to study the same turbulent effects in air ($v \approx 0.15$ cm^{2}s^{-1} at room temperature) than in an equivalent experiment in water ($v \approx 0.01$ cm^{2}s^{-1} at room temperature). Qualitatively, the Reynolds number describes the ratio of inertial forces $\mathbf{u} \cdot \nabla \mathbf{u}$ and viscous forces $v\Delta \mathbf{u}$. Accordingly, it plays a key role in understanding the onset of turbulence: for very low Reynolds numbers, for instance, the nonlinear terms are negligible which results in laminar fluid motion. However, an increase in Reynolds number results in a transition to a steady spatially structured flow, then to a spatially structured time-periodic flow, and then via a quasi-periodic transition to a flow that is chaotic in time [1]. In this monograph, we will not focus on this transition to turbulence but we are rather interested in studying the case of high Reynolds numbers. This regime of fully developed turbulence possesses considerable numbers of degrees of freedom, which leads to fundamentally new spatio-temporal behavior compared to the ones encountered in chaotic systems. In fact, it can be shown [2] that the number of degrees of freedom increases with the Reynolds number according to

$$\# \text{ degrees of freedom} \sim \mathrm{Re}^{9/4} \,. \tag{2.1.6}$$

For turbulent flows, it is therefore futile to describe individual time variations of *all* generalized coordinates and it may be appropriate to consider only ensembles of these generalized coordinates in a statistical sense, which will be further discussed in Chap. 3.

Apart from the complexity of the nonlinear term in the Navier-Stokes equation (2.1.1), another substantial influence on the spatio-temporal behavior of the solutions stems from the pressure term. For an incompressible flow, the pressure $p(\mathbf{x}, t)$ can be determined in taking the divergence of the Navier-Stokes equation (2.1.1) which yields

$$\Delta p(\mathbf{x}, t) = -\nabla \cdot [\mathbf{u}(\mathbf{x}, t) \cdot \nabla \mathbf{u}(\mathbf{x}, t)] . \tag{2.1.7}$$

Hence, the pressure $p(\mathbf{x}, t)$ is determined by a Poisson equation with a source term that is given by the divergence of the nonlinearity in Eq. (2.1.1). As it is discussed in Appendix 1, the solution of this equation in three dimensions reads

$$p(\mathbf{x}, t) = \frac{1}{4\pi} \int d\mathbf{x}' \frac{\nabla_{\mathbf{x}'} \cdot [\mathbf{u}(\mathbf{x}', t) \cdot \nabla_{\mathbf{x}'} \mathbf{u}(\mathbf{x}', t)]}{|\mathbf{x} - \mathbf{x}'|} . \tag{2.1.8}$$

In an incompressible fluid, the pressure term in the Navier-Stokes equation thus acts as an additional *nonlocality*, which transforms it into an integro-differential equation: the velocity field at a point $\mathbf{x}$ at time t is *instantaneously* related to the velocity field at all other points $\mathbf{x}'$. For a more pictorial description of the effects of nonlocality, let us assume that the source term of the Poisson equation (2.1.7) consists of an eddy (we will give a more precise definition of the notion of an eddy in Sect. 2.3) localized at point $\mathbf{x}'$. This eddy will then cause a disturbance of the velocity field at a point $\mathbf{x}$ in form of the pressure gradient which is proportional to $\frac{1}{|\mathbf{x}-\mathbf{x}'|^2}$. Although the pressure term adds to the complexity of the Navier-Stokes equation by introducing strong couplings between different scales [3], it is also known to possess regularizing effects on singular structures that are formed by the nonlinearity in the limit Re $\to \infty$ or $\nu \to 0$. Therefore, from a mathematical point of view, the role of pressure in the Navier-Stokes equation is deeply connected to the pending proof of existence and smoothness of global solutions that is considered as one of the seven "millennium problems" proclaimed by the Clay Mathematics Institute [4].

2.2   The Equation of Energy Balance and the Local Energy Dissipation Rate

In the preceding section, we got acquainted with the basic equations of hydrodynamic turbulence. This section is devoted to one of the key quantities of turbulent fluid motion, i.e., the local energy dissipation rate

$$\varepsilon(\mathbf{x}, t) = \frac{\nu}{2} \sum_{i,j} \left(\frac{\partial u_i(\mathbf{x}, t)}{\partial x_j} + \frac{\partial u_j(\mathbf{x}, t)}{\partial x_i} \right)^2 . \tag{2.2.1}$$

The importance of this quantity in the realm of a statistical description of turbulence cannot be overstated and will be further elaborated upon in Chap. 3. For the moment it suffices to say that, even in the vicinity of vanishing viscosity, the *averaged local energy dissipation rate* of a turbulent flow remains a finite quantity due to the steepening of velocity gradients in Eq. (2.2.1). Hence, the energy dissipation rate exhibits strong local fluctuations that are profoundly influenced by geometrical structures of the flow in the limit of high Reynolds numbers. Before we delve deeper into the properties of the local energy dissipation rate and their implications for a statistical theory in the next sections, we will derive an equation that contains the local energy dissipation rate directly from the Navier-Stokes equation (2.1.1). To this end, we scalar multiply Eq. (2.1.1) by $\mathbf{u}(\mathbf{x}, t)$ which yields

$$\frac{1}{2} \frac{\partial}{\partial t} \mathbf{u}^2(\mathbf{x}, t) + \nabla \cdot \left[\mathbf{u}(\mathbf{x}, t) \frac{\mathbf{u}(\mathbf{x}, t)^2}{2} + \mathbf{u}(\mathbf{x}, t) p(\mathbf{x}, t) \right]$$
$$= \nu \mathbf{u}(\mathbf{x}, t) \cdot \Delta \mathbf{u}(\mathbf{x}, t) + \mathbf{u}(\mathbf{x}, t) \cdot \mathbf{F}(\mathbf{x}, t) . \tag{2.2.2}$$

Here, we made use of the incompressibility of the velocity field in order to pull the divergence in front of the square brackets on the l.h.s. of Eq. (2.2.2). Furthermore, we included an external force $\mathbf{F}(\mathbf{x}, t)$ on the r.h.s. of the Navier-Stokes equation (2.1.1). The viscous contributions can be rewritten according to

$$\nu \mathbf{u}(\mathbf{x}, t) \cdot \Delta \mathbf{u}(\mathbf{x}, t) = \frac{\nu}{2} \Delta \mathbf{u}^2(\mathbf{x}, t) - \nu \sum_{i,j} \left(\frac{\partial u_i(\mathbf{x}, t)}{\partial x_j} \right)^2$$
$$= \frac{\nu}{2} \Delta \mathbf{u}^2(\mathbf{x}, t) - \frac{\nu}{2} \sum_{i,j} \left(\frac{\partial u_i(\mathbf{x}, t)}{\partial x_j} + \frac{\partial u_j(\mathbf{x}, t)}{\partial x_i} \right)^2$$
$$+ \nu \sum_{i,j} \left(\frac{\partial u_i(\mathbf{x}, t)}{\partial x_j} \frac{\partial u_j(\mathbf{x}, t)}{\partial x_i} \right)$$
$$= \frac{\nu}{2} \Delta \mathbf{u}^2(\mathbf{x}, t) - \varepsilon(\mathbf{x}, t) + \nu \nabla \cdot [\mathbf{u}(\mathbf{x}, t) \cdot \nabla \mathbf{u}(\mathbf{x}, t)] . \tag{2.2.3}$$

By virtue of this relation, Eq. (2.2.2) converts into a balance equation for the kinetic energy density $e_{kin} = \frac{\mathbf{u}^2}{2}$ which reads

$$\frac{\partial}{\partial t} e_{kin}(\mathbf{x}, t) + \nabla \cdot \mathbf{J}^{kin}(\mathbf{x}, t) = q(\mathbf{x}, t) , \tag{2.2.4}$$

where the energy flux density $\mathbf{J}^{kin}$ and the source term q are given by

$$\mathbf{J}^{kin}(\mathbf{x}, t) = \mathbf{u}(\mathbf{x}, t) \left[\frac{\mathbf{u}^2(\mathbf{x}, t)}{2} + p(\mathbf{x}, t)\right] - \frac{\nu}{2} \nabla \mathbf{u}^2(\mathbf{x}, t) - \nu \mathbf{u}(\mathbf{x}, t) \cdot \nabla \mathbf{u}(\mathbf{x}, t),$$

$$q(\mathbf{x}, t) = \mathbf{u}(\mathbf{x}, t) \cdot \mathbf{F}(\mathbf{x}, t) - \varepsilon(\mathbf{x}, t) . \tag{2.2.5}$$

The individual terms that enter the equation for the energy flux density can be interpreted as follows: the term $\mathbf{u}(\mathbf{x}, t)\frac{\mathbf{u}^2(\mathbf{x},t)}{2}$ can be considered as the kinetic energy that is transported through the surface S of a fluid volume V per unit time. The other term $\mathbf{u}(\mathbf{x}, t)p(\mathbf{x}, t)$ is the work per unit time that is done by the displacement caused by $\mathbf{u}(\mathbf{x}, t)$ against the pressure. The remaining terms are of viscous nature. Turning to the source term $q(\mathbf{x}, t)$, the external forcing results in a power density $\mathbf{u}(\mathbf{x}, t) \cdot \mathbf{F}(\mathbf{x}, t)$ whereas the local energy dissipation rate introduced in Eq. (2.2.1) denotes the dissipated kinetic energy per unit mass and unit time. The conservative form of Eq. (2.2.4) suggests that the kinetic energy defined by

$$E_{kin}(t) = \int_V d\mathbf{x} \, \frac{\mathbf{u}^2}{2} = \int_V d\mathbf{x} \, e_{kin}(\mathbf{x}, t) \tag{2.2.6}$$

is solely changed by the energy dissipation rate and the power density averaged over space according to

$$\dot{E}_{kin}(t) = \int_V d\mathbf{x} \, q(\mathbf{x}, t) , \tag{2.2.7}$$

where we made use of Gauss' theorem and the assumption that the flux vanishes at the boundaries of the fluid volume V which yields

$$\int_V d\mathbf{x} \, \nabla \cdot \mathbf{J}^{kin}(\mathbf{x}, t) = \int_S d\mathbf{a} \cdot \mathbf{J}^{kin}(\mathbf{x}, t) = 0 . \tag{2.2.8}$$

The energy dissipation rate characterizes the nonlinear transfer of energy among the various excited scales in turbulent flows, which will be further elaborated upon in Chap. 3.

2.3 The Vorticity Equation

A more vivid representation of the consequences of the Navier-Stokes equation in the context of its engendered vortical structures is acquired in defining the vorticity

$$\omega(\mathbf{x}, t) = \nabla \times \mathbf{u}(\mathbf{x}, t) . \tag{2.3.1}$$

Whereas the nonlocality of the pressure avoids local accumulation of velocity, vorticity cannot be created or destroyed within the fluid interior. It organizes in structures

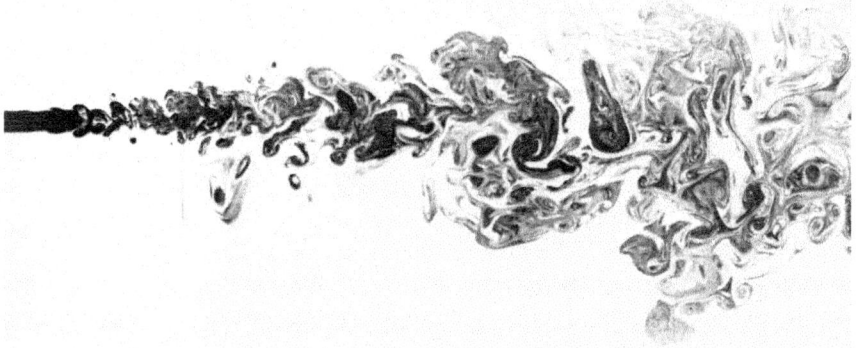

Fig. 2.1 Turbulent water jet visualized by laser-induced fluorescence [7]

such as vortex sheets or filaments and furthermore, in the case of two-dimensional fluid dynamics, it organizes itself into big clusters [5, 6]. The variety of vortical structures in a turbulent flow can be appreciated from fluid motion pictures like the one depicted in Fig. 2.1. The vorticity equation of fluid dynamics follows from the Navier-Stokes equation (2.1.1) according to

$$\frac{\partial}{\partial t}\boldsymbol{\omega}(\mathbf{x}, t) + \mathbf{u}(\mathbf{x}, t) \cdot \nabla \boldsymbol{\omega}(\mathbf{x}, t) = \boldsymbol{\omega}(\mathbf{x}, t) \cdot \nabla \mathbf{u}(\mathbf{x}, t) + \nu \Delta \boldsymbol{\omega}(\mathbf{x}, t) \ .$$

$$(2.3.2)$$

It is important to mention that although the pressure term dropped out of the vorticity equation, it still possesses a nonlocal character due to the velocity, which is related to the vorticity over the whole space by the Biot-Savart law

$$\mathbf{u}(\mathbf{x}, t) = \frac{1}{4\pi} \int d\mathbf{x}' \ \boldsymbol{\omega}(\mathbf{x}', t) \times \frac{\mathbf{x} - \mathbf{x}'}{|\mathbf{x} - \mathbf{x}'|^3} \ , \qquad (2.3.3)$$

derived in Appendix 1. The second term on the l.h.s. of the vorticity equation (2.3.2) implies advection of vorticity by the velocity field. The first term on the r.h.s., however, is the so-called *vortex stretching term*. This term leads to an amplification of the vorticity; it tends to concentrate the vorticity into thin vortex tubes or sheets on various scales. The vortex stretching term vanishes in two-dimensional flows which leads to the clustering of vorticity.

The vorticity equation in fluid mechanics (2.3.2) exhibits several exact solutions. For example, for inviscid fluid flows in two dimensions, the partial differential equation can be reduced to a Hamiltonian system of ordinary differential equations for the temporal evolution of so-called point vortices, provided that the vorticity initially consists of a superposition of delta distributions [8]. Another important vortex solution for the case of viscous three-dimensional flows, the Lamb-Oseen vortex, is a single azimuthal-symmetric vortex that reduces the vorticity equation to a simple heat equation. The vortex solution therefore decays in time due to dissipation. Contrary to latter solution, the so-called Burgers vortex is a stationary solution in the presence of dissipation, since it is exposed to an azimuthal strain field. For further discussion of vortex solutions and vortex methods, the reader is referred to [9, 10].

2.3.1 Kelvin's Circulation Theorem in Fluid Mechanics

An important consequence which follows from the vorticity equation of ideal fluid dynamics,

$$\frac{\partial}{\partial t}\boldsymbol{\omega}(\mathbf{x}, t) + \mathbf{u}(\mathbf{x}, t) \cdot \nabla \boldsymbol{\omega}(\mathbf{x}, t) = \boldsymbol{\omega}(\mathbf{x}, t) \cdot \nabla \mathbf{u}(\mathbf{x}, t) , \qquad (2.3.4)$$

is the conservation of the line integral of the velocity $\mathbf{u}$ along a closed curve C, which is known as the circulation

$$\Gamma = \oint_C d\mathbf{r} \cdot \mathbf{u} = \int_A d\mathbf{a} \cdot \boldsymbol{\omega} . \qquad (2.3.5)$$

In the last step, the line integral is written as a surface integral of the vorticity, according to Stokes' theorem. A is thereby a surface bounded by the closed curve C. Kelvin's circulation theorem now states that any circulation Γ about a closed curve that moves within an incompressible and inviscid fluid is a conserved quantity. If the fluid is subject to viscous forces, the circulation along the trajectories of particles moving with the fluid is no longer conserved.

2.3.2 Vortex Stretching and Vortical Structures

An explanation of the effects of vortex stretching in three-dimensional flows follows from the consideration of a vortex tube. It is commonly believed that dominant structures in three-dimensional turbulence consist of such vortex tubes or also vortex sheets. Figure 2.2 shows a schematic depiction of a vortex tube: The tube is being stretched if the longitudinal velocity u_l at A is smaller than at B. In this scenario, the contribution of the vortex stretching term,

$$\boldsymbol{\omega} \cdot \nabla u_l = |\boldsymbol{\omega}| \frac{du_l}{ds} , \qquad (2.3.6)$$

Fig. 2.2 Schematic depiction of a vortex tube. Stretching of the vortex tube leads to a positive vortex stretching term in Eq. (2.3.2) and, hence, an amplification of vorticity

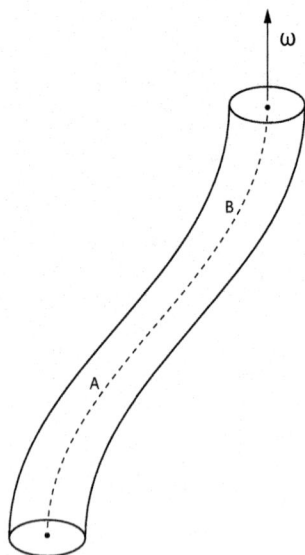

to the vorticity equation is positive since $du_l/ds > 0$ and thus results in an amplification of the latter. Here, ds is a line segment along the filament in Fig. 2.2. By contrast, compression of the vortex tube results in decreased values of the vorticity field.

A more formal treatment of the vortex stretching term in Eq. (2.3.2) starts from a decomposition of the tensor

$$\frac{\partial u_i}{\partial x_j} = S_{ij} + T_{ij} \,, \tag{2.3.7}$$

into a symmetric part

$$S_{ij} = \frac{1}{2} \left[\frac{\partial u_i}{\partial x_j} + \frac{\partial u_j}{\partial x_i} \right] \,, \tag{2.3.8}$$

and an antisymmetric part

$$T_{ij} = \frac{1}{2} \left[\frac{\partial u_i}{\partial x_j} - \frac{\partial u_j}{\partial x_i} \right] \,. \tag{2.3.9}$$

Here, the rate of strain S_{ij} is related to the rate of energy dissipation according to $\varepsilon = 2\nu S_{ij} S_{ij}$, where we assume summation over equal indices. The rate of strain at a point $\mathbf{x}$ within the fluid measures the rate at which distances of adjacent parcels of the fluid change with time in the vicinity of that point, which somewhat resembles the picture of vortex tube stretching from above. By contrast, the antisymmetric tensor $T_{ij} = -T_{ji}$ relates to the vorticity as follows:

$$\omega_i = -\epsilon_{ijk}T_{jk} \quad \text{or} \quad T_{ij} = -\frac{1}{2}\epsilon_{ijk}\omega_k , \tag{2.3.10}$$

where ϵ_{ijk} is the Levi-Civita tensor. Inserting Eq. (2.3.7) into the i-th component of the vortex stretching term yields

$$[\boldsymbol{\omega} \cdot \nabla \mathbf{u}]_i = \omega_j \frac{\partial u_i}{\partial x_j} = \omega_j(S_{ij} + T_{ij}) = S_{ij}\omega_j - \underbrace{\frac{1}{2}\epsilon_{ijk}\omega_j\omega_k}_{=0} = [S \cdot \boldsymbol{\omega}]_i . \tag{2.3.11}$$

We observe that the antisymmetric contribution to the vortex stretching term vanishes and latter solely describes the influence of strain rate on the vorticity. Inserting this result for the vortex stretching term into the vorticity equation (2.3.2) yields

$$\frac{\partial}{\partial t}\boldsymbol{\omega}(\mathbf{x}, t) + \mathbf{u}(\mathbf{x}, t) \cdot \nabla \boldsymbol{\omega}(\mathbf{x}, t) = S(\mathbf{x}, t) \cdot \boldsymbol{\omega}(\mathbf{x}, t) + \nu \Delta \boldsymbol{\omega}(\mathbf{x}, t) . \tag{2.3.12}$$

Hence, vorticity dynamics is inherently coupled to the strain rate tensor entering through the vortex stretching term. In this context, over the past two decades, numerical investigations have focused on the alignment of the vorticity with the eigenvectors of the strain rate tensor. As a matter of fact, the magnitude of the vortex stretching term in Eq. (2.3.12) depends on the strain rate eigenvalues s_i according to $|S_{ij}\omega_j| = \omega[s_i^2(\mathbf{e}_i \cdot \mathbf{e}_\omega)]$, where s_i are the eigenvalues and $\mathbf{e}_i$ the corresponding eigenvectors of the strain rate tensor [11].

Here, $s_1 \geq 0$ is the extensional eigenvalue, s_2 the intermediate eigenvalue, and $s_3 \leq 0$ the compressional eigenvalue, i.e., $s_1 \geq s_2 \geq s_3$. Moreover, in this representation, the strain rate tensor is trace-free, $s_1 + s_2 + s_3 = 0$, due to the incompressibility condition (2.1.2). In the example of the vortex tube depicted in Fig. 2.3a, we locally choose cylindrical coordinates at $\mathbf{x}$, which implies that the extensional direction lies in z-direction, i.e., $\mathbf{e}_1 = \mathbf{e}_z$, the compressional direction lies in radial direction, i.e., $\mathbf{e}_3 = \mathbf{e}_r$ and the intermediate direction lies in azimuthal direction, i.e., $\mathbf{e}_2 = \mathbf{e}_\varphi$. Seminal numerical investigations [12, 13] showed that vorticity preferentially aligns with the *intermediate eigenvector* $\mathbf{e}_2$. The physical mechanism that underlies this somehow counterintuitive finding is still up for debate. At first glance, Eq. (2.3.12) seems to suggest that, due to a positive maximum eigenvalue s_1, vorticity should preferentially align with the most extensional eigenvector. This deduction is based on the assumption that the rate of strain S decouples from the dynamics of $\boldsymbol{\omega}$. The Biot-Savart law (2.3.3), however, states that the velocity field is implicitly determined by the global vorticity, which leads to an intricate nonlinear behavior of vorticity dynamics in general. Here, the intermediate eigenvalue s_2 is a fluctuating quantity that can take on large negative or positive values (bounded by s_1 and s_3, respectively), but proves to be weakly positive in average.

The observed alignment of vorticity with the intermediate eigenvector $\mathbf{e}_2$ can be put in the context of vortex tube stretching as follows: As the vortex tube in Fig. 2.3a gets stretched in z-direction, vorticity tries to align itself with the intermediate eigen-

(a)   (b)   (c)

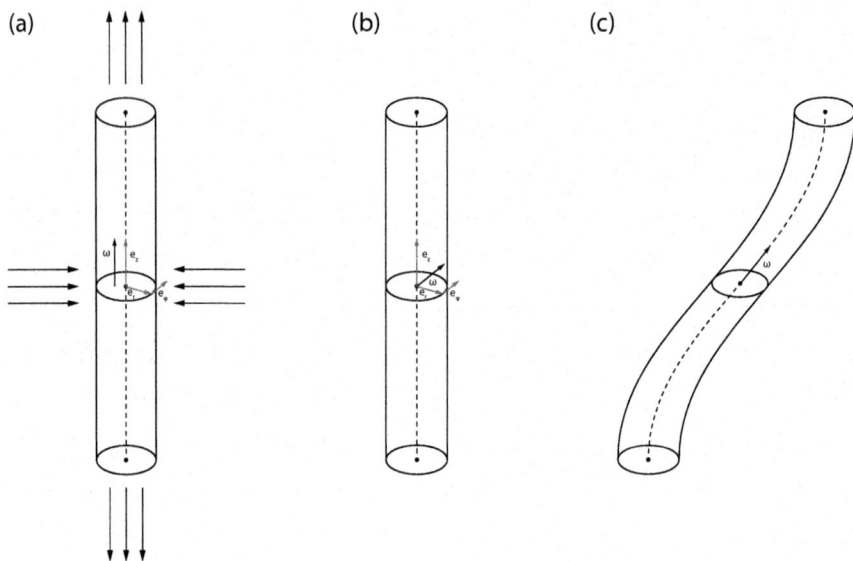

Fig. 2.3 **a** Schematic depiction of a vortex tube that gets stretched in z-direction and compressed in r-direction. Numerical simulations [12] show that the vorticity that initially points in z-direction tries to align itself with the intermediate eigenvector of the strain rate tensor $\mathbf{e}_\varphi$. **b** The vorticity has aligned itself with the intermediate eigenvector of the strain rate tensor that points in azimuthal direction $\mathbf{e}_\varphi$. The result **c** is a kinking that is also accompanied by a flattening of the tube into a sheet-like structure

vector $\mathbf{e}_\varphi$ in Fig. 2.3b. This results in kinking in combination with flattening of the vortex tube into a more sheet-like structure. Since above arguments entail vortex tubes as the most singular structures in a turbulent three-dimensional flow, at this point, we have to endorse at least some evidence that suggests their occurrence.

In this context, the increase of computational power opened up the way for direct numerical simulations (DNS) with increasing Reynolds number. Numerous scholars realized that the increase of Reynolds number is accompanied by complex structural changes that are strictly *non-self-similar*. Ishihara, Gotoh, and Kaneda [15], for instance, argued that small-scale turbulent fluctuations are dominated by elongated vortices similar to the schematically depicted vortex tube in Fig. 2.2. Such extreme vorticity events typically appear at a few times the mean or the standard deviation of the enstrophy $\Omega = \omega_i \omega_i$ (Greek for: "during the rotation"). Figure 2.4a, for instance, shows contour plots of enstrophy Ω (cyan) and energy dissipation rate ε (red) at a threshold of 10 times their mean value. These figures are taken from the currently largest DNS (8192^3 collocation points) by Yeung, Zhai, and Sreenivasan [14] that attain Taylor-Reynolds numbers (see Sect. 3.2.1.4 for an exact definition) of the order $Re_\lambda \approx 1300$. The contour plot at this rather low threshold shows sprinkled contours of Ω and ε. Increasing the threshold and zooming further into the sub-cube in Fig. 2.4b–c reveal more tube-like structures. Figure 2.4e solely shows enstrophy

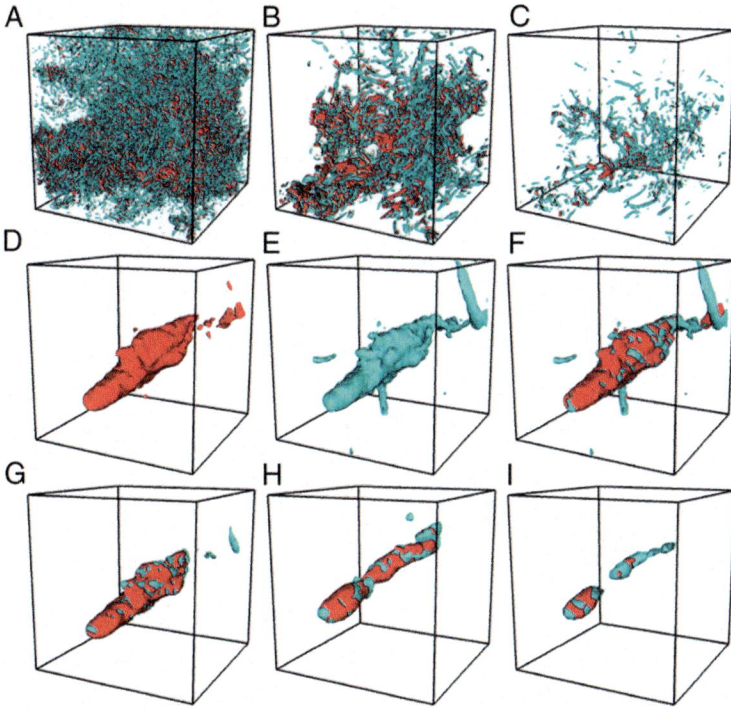

Fig. 2.4 Snapshot of three-dimensional contours of the enstrophy Ω (cyan) and the energy dissipation rate ε (red) taken from the currently largest DNS of 8192^3 grid points with Taylor-Reynolds number $\mathrm{Re}_\lambda \approx 1300$ by Yeung et al. [14]. Figures **a–i** correspond to different thresholds of the mean values and different subcubes of the 8192^3-cube. **a** 10, 768^3, **b** 30, 256^3, **c** 30, 256^3, **d** only energy dissipation rate is depicted for 300, 51^3, **e** only enstrophy is depicted for 300, 51^3, **f** 300, 51^3, **g** 600, 51^3, **h** 4800, 31^3 and **i** 96000, 31^3. The most intense vorticity events in (**i**) appear to be rather stocky in contrast to the tube-like structures of moderately large vorticity events (**d–f**)

and consists of a single vortex tube. Here, regions of high vorticity apparently tend to be wrapped around by sheets of intense dissipation, which can be deduced from Fig. 2.4d. More intense vorticity events observed by Yeung et al., however, are at odds with our current perception of vortex tubes as being the utmost singular structures in high Reynolds number turbulent flows. Tube-like structures, hence, correspond to moderately large vorticity events whereas the most intense vortical structures roughly correspond to rather stocky structures as the one depicted in Fig. 2.4i. Furthermore, the wrapping around of dissipation layers around these vorticity structures is not as pronounced as for moderate events. Figure 2.5 shows the probability density functions (PDFs) of the enstrophy (cyan) and the energy dissipation rate (red) from the simulations by Yeung et al. [14].

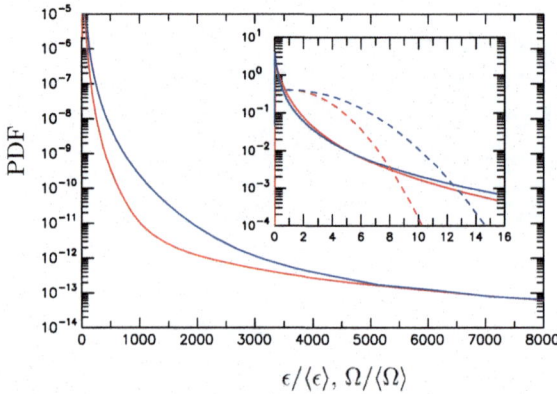

Fig. 2.5 Probability density functions (PDFs) of enstrophy Ω (cyan) and energy dissipation rate ε from the DNS of Yeung et al. [14]. Both enstrophy and energy dissipation rate PDF show pronounced non-Gaussian behavior that manifests itself by intense rare vorticity and energy dissipation events. The inset shows data for $0 - 16$ mean values that considerably deviates from Gaussian distributions with corresponding variances (dashed lines)

Due to the rare intense vorticity events observed in Fig. 2.4, PDFs exhibit strongly non-Gaussian behavior. Rare vorticity and energy dissipation events can be detected significantly for ≈ 10000 times the mean values of enstrophy $\langle \Omega \rangle$ and energy dissipation rate $\langle \varepsilon \rangle$, respectively.

Another unresolved issue that arises in the context of vortical structures and their corresponding singular behavior is the question of differences between longitudinal and transverse velocity increments. Here, the velocity increment is defined according to $\mathbf{v}(\mathbf{r}, \mathbf{x}, t) = \mathbf{u}(\mathbf{x} + \mathbf{r}, t) - \mathbf{u}(\mathbf{x}, t)$. The velocity increment can be decomposed into a longitudinal vector and a transverse vector with the help of the identity $\mathbf{e}_r \times (\mathbf{e}_r \times \mathbf{v}) = \mathbf{e}_r (\mathbf{e}_r \cdot \mathbf{v}) - \mathbf{v}$. The decomposition leads to the definition of the longitudinal velocity increment

$$\mathbf{v}_r(\mathbf{r}, \mathbf{x}, t) = \mathbf{e}_r \left([\mathbf{u}(\mathbf{x} + \mathbf{r}, t) - \mathbf{u}(\mathbf{x}, t)] \cdot \mathbf{e}_r \right) , \qquad (2.3.13)$$

and the transverse velocity increment

$$\mathbf{v}_t(\mathbf{r}, \mathbf{x}, t) = -\mathbf{e}_r \times (\mathbf{e}_r \times [\mathbf{u}(\mathbf{x} + \mathbf{r}, t) - \mathbf{u}(\mathbf{x}, t)]) . \qquad (2.3.14)$$

Numerical as well as experimental observations point at different statistical behavior of the longitudinal and transverse increments [16]. More specifically, transverse velocity increment statistics appears to be more sensitive to the influence of intense

singular structures than the longitudinal one. A possible explanation for the cause of
this persistent different statistical behavior, once more, was based on rough geometrical considerations of the underlying singular structures: under the assumption, that
the longitudinal direction points in the direction of the vortex tube (dashed line) in
Fig. 2.2, transverse velocity increments should experience the Biot-Savart singularity
from crossing the vortex tube. Therefore, the geometrical argument seems to explain
why transverse increments possess different statistical behavior than longitudinal
increments. However, recent numerical studies of magneto-hydrodynamic turbulence [17] showed that this rough geometrical picture is not an adequate description
for the longitudinal-transverse velocity increment differences: Despite the fact that
magneto-hydrodynamics exhibits even more pronounced singular structures (such
as vortex sheets or current sheets), no substantial differences between longitudinal
and transverse increment statistics could be observed in DNS of three-dimensional
magneto-hydrodynamic turbulence. This rather counterintuitive finding could be
attributed to regions of depleted pressure that are the result of preferential local
alignment between velocity and magnetic field. Apparently, the pressure term, in its
nonlocal representation (2.1.8), plays a key role for different increment statistics in
hydrodynamics. A similar alignment mechanism might also lead to depleted pressure
regions in hydrodynamic turbulence, although these regions are not encountered as
frequently as in the case of magneto-hydrodynamic turbulence.

2.4 Burgers Equation: A Simplistic Version of Hydrodynamic Turbulence

The preceding sections summarized the overwhelming complexity and mathematical
problems that arise from the Navier-Stokes equation. In 1939, the Dutch scientist
J. M. Burgers [18] proposed a simplified version of the Navier-Stokes equation in the
hope that it still preserves the main signatures of turbulence. In dropping the pressure
term in Eq. (2.1.1) and restricting ourselves to one dimension, we obtain the Burgers
equation

$$\frac{\partial}{\partial t}u(x,t) + u(x,t)\frac{\partial}{\partial x}u(x,t) = v\frac{\partial^2}{\partial x^2}u(x,t) . \tag{2.4.1}$$

The similarities of this equation to the Navier-Stokes equation are as interesting as its differences. First of all, the nonlinearity conserves $\int dx u(x,t)$ as well
as $\int dx u(x,t)^2$, which is also true for the Navier-Stokes equation. Moreover, the
interplay between the nonlinearity and the viscous term results in the steepening of
velocity gradients by the nonlinear term until it is then subject to increased energy
dissipation due to the viscous term. Therefore, both the Navier-Stokes equation and

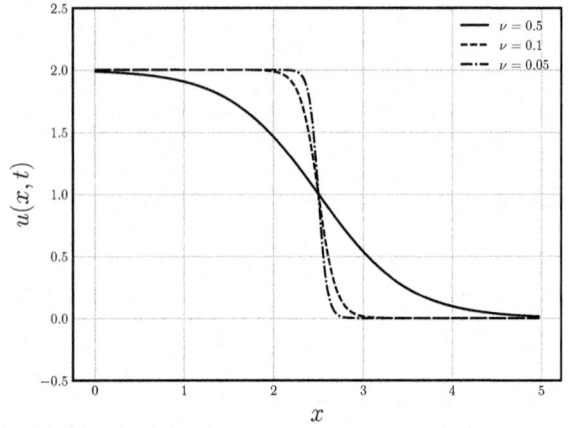

Fig. 2.6 Shock solution of Burgers equation (2.4.2) for $x_c = 2$ and $t = 0$. The three shock profiles correspond to different viscosities ν

the Burgers equation imply a transfer of energy from the large-scale motion to smaller scales. However, the main difference is that, due to the compressibility of the velocity field in Eq. (2.4.1), Burgers turbulence at high Reynolds numbers is dominated by shock structures. A simple shock solution of Eq. (2.4.1), for instance, reads

$$u(x, t) = 1 - \tanh\left(\frac{x - x_c - t}{2\nu}\right) .$$

(2.4.2)

The velocity profile of such a shock is depicted in Fig. 2.6. For this case, the averaged energy dissipation rate can be calculated explicitly according to

$$\langle \varepsilon \rangle = 2\nu \int_{-\infty}^{\infty} dx \left(\frac{\partial u(x, t)}{\partial x}\right)^2 = \frac{4}{3} ,$$

(2.4.3)

and is therefore independent of the viscosity. The shocks tend to coalesce and thus decrease the complexity of an initially complicated velocity field. By contrast, the Navier-Stokes equation leads to a nearly infinite hierarchy of instabilities and thus increases complexity over time. In fact, it was shown by Hopf [19] and Cole [20] that the Burgers equation is integrable which can be established in using the so-called Hopf-Cole transformation

$$u(x, t) = -2\nu \frac{\partial}{\partial x} \ln \psi(x, t) .$$

(2.4.4)

Via this transformation, the nonlinear Burgers equation can be turned into the linear heat equation

$$\frac{\partial}{\partial t}\psi(x,t) = \nu\frac{\partial^2}{\partial x^2}\psi(x,t) , \qquad (2.4.5)$$

which has an exact solution.

Another important aspect of turbulence that can be exemplified with the Burgers equation is the notion of the so-called *dissipation anomaly* which will be further discussed in Sects. 5.1 and 5.4. To this end, let us multiply the inviscid Burgers equation with $u(x,t)$ and integrate from 0 to a length L, e.g., the box length in Fig. 2.6. We obtain

$$\frac{\partial}{\partial t}E_{kin}(t) = \frac{1}{2}\frac{\partial}{\partial t}\int_0^L dx\, u^2(x,t) = -\frac{1}{3}\int_0^L dx\, \frac{\partial}{\partial x}u^3(x,t) . \qquad (2.4.6)$$

Under the assumption that the velocity field is smooth and that we have periodic boundary conditions, i.e., $u(x=0,t) = u(x=L,t)$, the term on the r.h.s. vanishes and we obtain $\frac{\partial}{\partial t}E_{kin}(t) = 0$. This is nothing else than the conservation of kinetic energy by the inviscid Burgers equation. However, if the velocity field develops a shock at x_0, we have to integrate around that shock (otherwise, we would not be able to change differentiation and integration as we did in Eq. (2.4.6))

$$\begin{aligned}
\frac{\partial}{\partial t}E_{kin}(t) &= \frac{1}{2}\frac{\partial}{\partial t}\int_0^L dx\, u^2(x,t) \\
&= -\frac{1}{3}\lim_{\epsilon\to 0}\left[\int_0^{x_0-\epsilon} dx\, \frac{\partial}{\partial x}u^3(x,t) + \int_{x_0+\epsilon}^L dx\, \frac{\partial}{\partial x}u^3(x,t)\right] \\
&= \frac{1}{3}\lim_{\epsilon\to 0}\left[u^3(x_0+\epsilon,t) - u^3(x_0-\epsilon,t)\right] , \qquad (2.4.7)
\end{aligned}$$

which leads to the conclusion $\frac{\partial}{\partial t}E_{kin}(t) < 0$. This implies that, even in the absence of viscosity, the kinetic energy is not conserved; it is known as dissipation anomaly. In this context, a remarkable conjecture for three-dimensional turbulence has been established by Onsager [21] (see also the review by Eyink and Sreenivasan [22]). Onsager concludes that "turbulent dissipation as described could take place just as readily without the final assistance by viscosity. In the absence of viscosity, the standard proof of the conservation of energy does not apply, because the velocity field does not remain differentiable!". He further states that in such "ideal turbulence", the velocity field cannot obey a Hölder condition of the form

$$|\mathbf{u}(\mathbf{x}+\mathbf{r}) - \mathbf{u}(\mathbf{x})| < Cr^h , \qquad (2.4.8)$$

for $h > 1/3$ and some $C > 0$. Otherwise, the kinetic energy is a conserved quantity. The conjecture can be summarized as follows:

Onsager's conjecture:

If the velocity field in ideal three-dimensional turbulence obeys a Hölder condition of the form

$$|\mathbf{u}(\mathbf{x} + \mathbf{r}) - \mathbf{u}(\mathbf{x})| < Cr^h ,\qquad(2.4.9)$$

for some $C > 0$, then

$$\frac{\partial}{\partial t} E_{kin}(t) = 0 \quad \text{for} \quad h > \frac{1}{3} \quad \text{(smooth velocity field)} , \quad(2.4.10)$$

$$\frac{\partial}{\partial t} E_{kin}(t) < 0 \quad \text{for} \quad h \leq \frac{1}{3} \quad \text{(rough velocity field)} . \quad(2.4.11)$$

It has to be stressed that, whereas Eq. (2.4.10) proves to be correct, Eq. (2.4.11) still remains a conjecture [23]. Onsager's conjecture already predicts an upper bound for the scaling of velocity increments in high Reynolds number turbulence. The latter prediction will come up repeatedly in the next chapters that are devoted to statistical descriptions of hydrodynamic turbulence.

Appendix 1: The Biot-Savart Law

Due to the incompressibility condition (2.1.2), we can express the velocity field $\mathbf{u}(\mathbf{x}, t)$ as the sum of a gradient field and the curl of a vector potential $\mathbf{A}(\mathbf{x}, t)$ according to

$$\mathbf{u}(\mathbf{x}, t) = \nabla\phi(\mathbf{x}, t) + \nabla \times \mathbf{A}(\mathbf{x}, t) ,\qquad(2.4.12)$$

where the scalar field $\phi(\mathbf{x}, t)$ satisfies the Laplace equation

$$\Delta\phi(\mathbf{x}, t) = 0 .\qquad(2.4.13)$$

Moreover, we can make use of the relation $\nabla \times [\nabla \times \mathbf{A}(\mathbf{x}, t)] = \nabla(\nabla \cdot \mathbf{A}(\mathbf{x}, t)) - \Delta\mathbf{A}(\mathbf{x}, t) = \omega(\mathbf{x}, t)$. Since we possess a certain liberty in the choice of the vector potential, we can assume that it is incompressible and obtain a Poisson equation of the form

$$\Delta\mathbf{A}(\mathbf{x}, t) = -\omega(\mathbf{x}, t) .\qquad(2.4.14)$$

We can formally solve this equation by making use of Green's function of the Laplacian $G(\mathbf{x} - \mathbf{x}')$, which satisfies

$$\Delta G(\mathbf{x} - \mathbf{x}') = -\delta(\mathbf{x} - \mathbf{x}') .\qquad(2.4.15)$$

Hence, the integrated form of Eq. (2.4.14) reads

$$\mathbf{A}(\mathbf{x}, t) = \int d\mathbf{x}' G(\mathbf{x} - \mathbf{x}')\boldsymbol{\omega}(\mathbf{x}', t) . \qquad (2.4.16)$$

Green's functions of the Laplacian in an infinite two-, respectively, three-dimensional space read

$$G(\mathbf{x} - \mathbf{x}') = \begin{cases} -\frac{1}{2\pi} \ln |\mathbf{x} - \mathbf{x}'| & \text{for } \mathbf{x} \in \mathbb{R}^2, \\[2mm] \frac{1}{4\pi} \frac{1}{|\mathbf{x} - \mathbf{x}'|} & \text{for } \mathbf{x} \in \mathbb{R}^3 . \end{cases} \qquad (2.4.17)$$

The velocity field in Eq. (2.4.12) thus reads

$$\mathbf{u}(\mathbf{x}, t) = \nabla_{\mathbf{x}}\phi(\mathbf{x}, t) + \nabla_{\mathbf{x}} \times \int d\mathbf{x}' G(\mathbf{x} - \mathbf{x}')\boldsymbol{\omega}(\mathbf{x}', t) . \qquad (2.4.18)$$

In the absence of a gradient field, we can express the velocity field by Biot-Savart law as

$$\mathbf{u}(\mathbf{x}, t) = \int d\mathbf{x}' \, \boldsymbol{\omega}(\mathbf{x}', t) \times \mathbf{K}(\mathbf{x} - \mathbf{x}') , \qquad (2.4.19)$$

where we have introduced the quantity

$$\mathbf{K}(\mathbf{x} - \mathbf{x}') = -\nabla_{\mathbf{x}} G(\mathbf{x} - \mathbf{x}') . \qquad (2.4.20)$$

According to Eq. (2.4.17), we obtain

$$\mathbf{K}(\mathbf{x} - \mathbf{x}') = \begin{cases} \frac{1}{2\pi} \frac{\mathbf{x} - \mathbf{x}'}{|\mathbf{x} - \mathbf{x}'|^2} & \text{for } \mathbf{x} \in \mathbb{R}^2 , \\[2mm] \frac{1}{4\pi} \frac{\mathbf{x} - \mathbf{x}'}{|\mathbf{x} - \mathbf{x}'|^3} & \text{for } \mathbf{x} \in \mathbb{R}^3 . \end{cases} \qquad (2.4.21)$$

References

1. Nelkin, M.: In what sense is turbulence an unsolved problem? Science **255**(5044), 566–70 (1992)
2. Landau, L.D., Lifshitz, E.M.: Statistical Physics, Third Edition: Volume 5 (Course of Theoretical Physics). Butterworth-Heinemann (1987)
3. Gotoh, T., Nakano, T.: Role of pressure in turbulence. J. Stat. Phys. **113**(5–6), 855–874 (2003)
4. Fefferman, C.L.: Existence and smoothness of the Navier-Stokes equation. Millenn. Prize Probl. **1**, 1–5 (2000)
5. Kraichnan, R.H.: Inertial Ranges in Two-Dimensional Turbulence. Phys. Fluids **10**(7), (1967)
6. Friedrich, J., Friedrich, R.: Generalized vortex model for the inverse cascade of two-dimensional turbulence. Phys. Rev. E **88**(5), 1–5 (2013)
7. Van Dyke, M.: An Album of Fluid Motion. Parabolic Press, Inc., 14 edition (2012)
8. Aref, H.: Point vortex dynamics: A classical mathematics playground. J. Math. Phys. **48**(6), (2007)

9. Faust, G., Argyris, J., Haase, M., Friedrich, R.: An Exploration of Dynamical Systems and Chaos. Springer (2015)
10. Saffman, P.G.: Vortex Dynamics. Cambridge University Press (1993)
11. Hamlington, P.E., Schumacher, J., Dahm, W.J.A.: Direct assessment of vorticity alignment with local and nonlocal strain rates in turbulent flows. Phys. Fluids **20**(11), 111703 (2008)
12. Ashurst, W.T., Kerstein, A.R., Kerr, R.M., Gibson, C.H.: Alignment of vorticity and scalar gradient with strain rate in simulated Navier-Stokes turbulence. Phys. Fluids **30**(8), 2343–2353 (1987)
13. Siggia, E.D.: Numerical study of small-scale intermittency in three-dimensional turbulence. J. Fluid Mech. **107**, 375–406 (1981)
14. Yeung, P.K., Zhai, X.M., Sreenivasan, K.R.: Extreme events in computational turbulence. Proc. Natl. Acad. Sci. **112**(41), 12633–12638 (2015)
15. Ishihara, T., Gotoh, T., Kaneda, Y.: Study of high-Reynolds Number isotropic turbulence by direct numerical simulation. Annu. Rev, Fluid Mech (2009)
16. Grauer, R., Homann, H., Pinton, J.-F.: Longitudinal and transverse structure functions in high-Reynolds-number turbulence. New J. Phys. **14**, 63016 (2012)
17. Friedrich, J., Homann, H., Schäfer, T., Grauer, R.: Longitudinal and transverse structure functions in high Reynolds-number magneto-hydrodynamic turbulence. New J. Phys. **18**(12), 125008 (2016)
18. Burgers, J.M.: The Nonlinear Diffusion Equation. Reidel, Dordrecht (1974)
19. Hopf, E.: The partial differential equation ut + uux = μxx. Commun. Pure Appl. Math. **3**(3), 201–230 (1950)
20. Cole, J.D.: On a quasi-linear parabolic equation occuring in aerodynamics. Q. Appl. Math. **9**(3), 225–236 (1951)
21. Onsager, L.: Statistical hydrodynamics. Nuovo Cim. **6**(2), 279–287 (1949)
22. Eyink, G.L., Sreenivasan, K.R.: Onsager and the theory of hydrodynamic turbulence. Rev. Mod. Phys. **78**(1), 87–135 (2006)
23. Constantin, P., E, W., Titi, E.S.: Onsager's conjecture on the energy conservation for solutions of Euler's equation. Comm. Math. Phys. 207–209 (1994)

Chapter 3
Statistical Formulation of the Problem of Turbulence

Chapter 2 already highlighted the fact that a turbulent flow is a mechanical nonlinear system with a very large number of degrees of freedom, which can be estimated as a function of the Reynolds number according to $Re^{9/4}$. Therefore, in contrast to laminar fluid motion where only a few degrees of freedom are excited [1], it is nearly impossible to describe individual time variations of all generalized coordinates (velocity $\mathbf{u}$, pressure p, or temperature T) in a fully developed turbulent flow accurately. It is thus suggested to consider only *ensembles* of these generalized coordinates. Accordingly, successful approaches to hydrodynamic turbulence have to involve a treatment of the deterministic equations of fluid mechanics via statistical and stochastic methods.

The latter field of fluid mechanics is also referred to as *statistical hydrodynamics* and is inseparably associated with the names of Onsager [2], Kolmogorov [3], von Weiszäcker [4], and Heisenberg [5]. This chapter is organized as follows: first, we analyze the notion of statistical averages and certain symmetries that lead to an invariant theory of these statistical quantities. After these definitions, we point out an important phenomenological description suggested by Richardson and Kolmogorov. Here, the large number of interlacing eddies is conceived as a hierarchical ordering of eddies that exchange energy from large to small scales where kinetic energy is transformed into heat due to viscous forces. Using the concept of the *turbulent energy cascade*, Kolmogorov [3] was able to make certain quantitative statements for the energy transfer rate as well as for characteristic scales of turbulent fluid motion. After the determination of these characteristic scales, we derive evolution equations for correlation functions as well as for probability density functions directly from the Navier-Stokes equation (2.1.1). An inherent difficulty of these evolution equations is the so-called *closure problem of turbulence* that leads to unclosed terms in the evolution equation of a certain order. Before we present methods that were devised to directly overcome the closure problem in Chap. 4, we discuss other phenomenological models of turbulence in the present chapter. The development of latter models was profoundly influenced by the phenomenon of *intermittency*, which can undoubtedly

© Springer Nature Switzerland AG 2021

J. Friedrich, *Non-perturbative Methods in Statistical Descriptions of Turbulence*,
Progress in Turbulence - Fundamentals and Applications 1,
https://doi.org/10.1007/978-3-030-51977-3_3

be considered as one of the key signatures of turbulent fluid motion. Here, deviations from Kolmogorov's *self-similar* theory (K41) manifest themselves through strong small-scale fluctuations of velocity field increments.

3.1   Foundations of Statistical Hydrodynamics

In the context of statistical hydrodynamics, the velocity field in a turbulent flow is assumed to take on random values $\mathbf{u}(\mathbf{x}, t)$. Consequently, we are mainly interested in mean values of this random function: Since we know that it is impossible to determine the exact value of $\mathbf{u}(\mathbf{x}, t)$, we rather assume that the random values of $\mathbf{u}(\mathbf{x}, t)$ are distributed according to certain probability laws. Henceforward, the method of taking averages consists in finding an appropriate statistical ensemble. This takes into account the totality of all possible realizations of a turbulent flow, which only differ in their initial conditions at all points $\mathbf{x}$ at time t. The ensemble is characterized by a statistical distribution for the initial field and we can derive a probability density function (PDF)

$$g_1(\mathbf{u}_1, \mathbf{x}_1, t) = \langle \delta(\mathbf{u}_1 - \mathbf{u}(\mathbf{x}_1, t)) \rangle \, , \tag{3.1.1}$$

for the velocity field $\mathbf{u}_1$ at position $\mathbf{x}_1$ and time t. In general, there will be a statistical connection between several velocities $\mathbf{u}_i$ at points $\mathbf{x}_i$, and we have to consider the n-point PDF

$$g_n(\mathbf{u}_n, \mathbf{x}_n; \mathbf{u}_{n-1}, \mathbf{x}_{n-1}; \ldots; \mathbf{u}_1, \mathbf{x}_1, t) = \prod_{i=1}^{n} \langle \delta(\mathbf{u}_i - \mathbf{u}(\mathbf{x}_i, t)) \rangle \, , \tag{3.1.2}$$

and then perform the continuum limit. In order to take an average of some function $F(\mathbf{u}_n, \ldots, \mathbf{u}_1, t)$ at a certain instance in time, an integration over all possible realizations $\mathbf{u}_i$ with a weight of the joint PDF (3.1.2) has to be performed according to

$$\langle F \rangle_{ensemble} = \int d\mathbf{u}_n \ldots d\mathbf{u}_1 \, F(\mathbf{u}_n, \ldots, \mathbf{u}_1, t) g_n(\mathbf{u}_n, \mathbf{x}_n; \ldots; \mathbf{u}_1, \mathbf{x}_1, t) \, . \tag{3.1.3}$$

This rather mathematical method of taking averages is not useful for experimental applications, since the totality of all realizations is never accessible in experiments or direct numerical simulations. Therefore, we invoke the *ergodicity hypothesis*, which states that the ensemble average can be replaced by temporal averages

$$\langle F \rangle_{temporal} = \langle F(t) \rangle = \lim_{T \to \infty} \frac{1}{T} \int_0^T dt' \, F(t + t') \, . \tag{3.1.4}$$

The PDFs provide an adequate mean to characterize a turbulent flow with respect to the following aspects:

- **Homogeneity**:
The multi-point distributions are functions of the relative distances $\mathbf{r}_{ij} = \mathbf{x}_i - \mathbf{x}_j$ only. This can be interpreted as the invariance of the PDFs under translations $\mathbf{X}$

$$g_n(\mathbf{u}_n, \mathbf{x}_n + \mathbf{X}; \ldots; \mathbf{u}_1, \mathbf{x}_1 + \mathbf{X}, t) = g_n(\mathbf{u}_n, \mathbf{x}_n; \ldots; \mathbf{u}_1, \mathbf{x}_1, t) . \qquad (3.1.5)$$

Take, for instance, the two-point velocity field correlation function

$$C_{ij}(\mathbf{x}, \mathbf{x}', t) = \langle u_i(\mathbf{x}, t) u_j(\mathbf{x}', t) \rangle . \qquad (3.1.6)$$

Under the assumption of homogeneity, it solely depends on the relative distance $\mathbf{r} = \mathbf{x} - \mathbf{x}'$ between the velocity field $u_i(\mathbf{x}, t)$ at point $\mathbf{x}$ and the velocity field $u_j(\mathbf{x}', t)$ at point $\mathbf{x}'$, or

$$C_{ij}(\mathbf{x}, \mathbf{x}', t) = C_{ij}(\mathbf{r}, t) . \qquad (3.1.7)$$

- **Stationarity**:
The PDF is independent of time t, and hence it is invariant under time translations τ

$$g_n(\mathbf{u}_n, \mathbf{x}_n; \ldots; \mathbf{u}_1, \mathbf{x}_1, t + \tau) = g_n(\mathbf{u}_n, \mathbf{x}_n; \ldots; \mathbf{u}_1, \mathbf{x}_1, t) . \qquad (3.1.8)$$

- **Isotropy**:
The PDF is invariant under rotations $R \in SO(3)$ with respect to an arbitrary axis of the coordinate system

$$g_n(R\mathbf{u}_n, R\mathbf{x}_n; \ldots; R\mathbf{u}_1, R\mathbf{x}_1, t) = g_n(\mathbf{u}_n, \mathbf{x}_n; \ldots; \mathbf{u}_1, \mathbf{x}_1, t) . \qquad (3.1.9)$$

This invariance is broken in the presence of buoyancy and rotational forces, which suppress vertical and horizontal motion. The assumption of isotropy and homogeneity also has a great influence on the tensorial form of the two-point velocity correlation function $C_{ij}(\mathbf{r}, t)$. For instance, it follows from the calculus of isotropic tensors discussed in Appendix 1 that its tensorial form is given according to

$$C_{ij}(\mathbf{r}, t) = (C_{rr}(r, t) - C_{tt}(r, t)) \frac{r_i r_j}{r^2} + C_{tt}(r, t) \delta_{ij} , \qquad (3.1.10)$$

where $C_{rr}(r, t)$ and $C_{tt}(r, t)$ are the longitudinal and transverse correlation functions, with respect to the relative distance $\mathbf{r}$. Finally, we want to mention that some quantities such as $\langle u_i(\mathbf{x}, t) \omega_j(\mathbf{x}', t) \rangle$ are not invariant under the full rotation group, i.e., they lack mirror symmetry [6–8].

3.2 Phenomenology of the Turbulent Energy Cascade

Perhaps one of the seminal works on the concept of the turbulent energy cascade is by Richardson [9]. Richardson's cascade model is based on the idea that turbulent flows are governed by vortex elements of varying length scales that obey a certain *hierarchical ordering*. The occurrence of vortices of a variety of length scales, however, has already been documented by Leonardo da Vinci in his famous sketch books. A typical drawing of da Vinci is shown in Fig. 3.1, the drawing of a flood. Astonishingly, da Vinci not only gave an accurate account of the interlacing vortex structures, but was also aware of the random fluid motion and its connection to the mean flow. As a matter of fact, he regularly complemented his drawings with comments such as [10]:

> Observe the motion of the surface of the water, which resembles that of hair, which has two motions, of which one is caused by the weight of the hair, the other by the direction of the curls, thus the water has eddying motions, one part of which is due to the principal current, the other to the random and reverse motion.

Nevertheless, a sound physical picture of the turbulent energy cascade has only been established by Richardson: if energy is injected at large scales L of the system,

Fig. 3.1 Leonardo da Vinci's drawing of a flood reveals his profound appreciation of the interlacing vortical structures of different length scales

large-scale vortical structures are generated. These structures, however, are subject to the vortex stretching mechanism discussed in Sect. 2.3.2 and decay after a short period of time. The central assumption in Richardson's cascade model is that the decay of large-scale vortical structures is accompanied by a transfer of kinetic energy to the next generation of vortices of smaller size, without dissipation of kinetic energy. This cascade process ultimately stops at a scale where viscous forces are comparable to the nonlinear energy transfer and energy is dissipated into heat.

Kolmogorov's famous K41 phenomenology [3] adopts the cascade model by Richardson and refines it with certain hypothesis.

Kolmogorov's first hypothesis of local isotropy:

The influence of anisotropies and boundary conditions is lost during the decay of the largest vortical structures. This implies that, in the case of high Reynolds numbers, every turbulent flow is statistically homogeneous and isotropic at scales $r \ll L$. A high Reynolds number turbulent flow thus possesses *universal character* at scales r that are smaller than scale L at which energy is injected into the system.

Accordingly, at scales $r \ll L$, kinetic energy is transferred to smaller scales until all kinetic energy is finally dissipated, which directly leads to

Kolmogorov's first similarity hypothesis:

Physical processes that occur at scales $r \ll L$ only depend on the averaged local energy dissipation rate $\langle \varepsilon \rangle$ (see Sect. 2.2) and the viscosity ν.

The first similarity hypothesis indicates the existence of a smallest characteristic scale η, the Kolmogorov microscale which will be the subject of Sect. 3.2.1.3. Intermediate scales, i.e., $\eta \ll r \ll L$ are captured by

Kolmogorov's second similarity hypothesis:

At sufficiently high Reynolds numbers, an *inertial range* of scales $\eta \ll r \ll L$ occurs, at which statistical quantities possess universal character, i.e., they do not depend on ν but only on $\langle \varepsilon \rangle$.

With the help of these hypothesis, Kolmogorov was able to derive quantitative predictions for certain statistical quantities in the inertial range. Before we describe his procedure in more detail, however, we will define the characteristic turbulent length scales which were already mentioned in the present section.

3.2.1   Characteristic Scales of Turbulent Fluid Motion

3.2.1.1   Taylor's Hypothesis

Although typical phenomenological descriptions of turbulence are usually based on spatial velocity increments $\mathbf{u}(\mathbf{x}', t) - \mathbf{u}(\mathbf{x}, t)$, these increments are often not accessible in experiments and velocity increments are rather measured in temporal domain. Nevertheless, under appropriate conditions, the obtained temporal increments measured at the time interval Δt can be transformed into spatial increments $\mathbf{r} = \mathbf{x} - \mathbf{x}'$ by use of Taylor's hypothesis.

Let us consider a wind tunnel experiment where the velocity $\mathbf{u}'(\mathbf{x}, t)$ is measured in the frame of reference of the mean flow, which points in the direction of $\mathbf{x}$. In the laboratory frame of reference, we would therefore measure the velocity

$$\mathbf{u}(\mathbf{x}, t) = \mathbf{u}'(\mathbf{x} - \mathbf{U}t, t) + \mathbf{U} ,\tag{3.2.1}$$

where $\mathbf{U}$ denotes the mean flow $\mathbf{U} = \langle \mathbf{u}(\mathbf{x}, t) \rangle$. Under the assumption that the mean flow is large compared to the fluctuations

$$\frac{\langle |\mathbf{u}'|^2 \rangle}{|\mathbf{U}|^2} \ll 1 ,\tag{3.2.2}$$

most of the time dependence in $\mathbf{u}(\mathbf{x}, t)$ arises from $\mathbf{x} - \mathbf{U}t$ in $\mathbf{u}'(\mathbf{x}, t)$. Therefore, by virtue of

$$\mathbf{r} = \mathbf{U}\Delta t ,\tag{3.2.3}$$

temporal variations of $\mathbf{u}(\mathbf{x}, t)$ can be interpreted as spatial variations of $\mathbf{u}'(\mathbf{x}, t)$.

3.2.1.2   The Integral Length Scale

The integral length scale L can roughly be considered as the typical length scale of the biggest vortical structures in the flow. It likewise corresponds to the scale at which energy is injected into the system. A more precise definition is based on the correlation between the velocity field at a point x and a point $x + r$

$$C(r) = \langle u(x + r)u(x) \rangle .\tag{3.2.4}$$

Although, here, we restrict ourselves to a one-dimensional velocity field, a generalization to higher dimensions and anisotropic flows is straightforward [11]. The integral length scale can now be defined as the scale in an isotropic and homogeneous flow at which the velocity field at two given points x and $x + r$ becomes statistically uncorrelated. Consequently, if one assumes an exponential decay of correlations in Eq. (3.2.4), i.e., $C(r) \approx C(0)e^{-r/L}$ for large r, we can define the integral length scale according to

$$L = \int_0^\infty dr \, \frac{C(r)}{C(0)} \, . \tag{3.2.5}$$

Henceforward, turbulent fluctuations that exceed the integral length scale L are statistically independent. Moreover, the integral length scale solely depends on the kinetic energy E_{kin} and the energy dissipation rate $\langle \epsilon \rangle$, $L = L(E_{kin}, \epsilon)$. By dimensional analysis, we find

$$L = \frac{E_{kin}^{\frac{3}{2}}}{\langle \varepsilon \rangle} = \frac{(\frac{1}{2} u_{rms}^2)^{\frac{3}{2}}}{\langle \varepsilon \rangle} \, , \tag{3.2.6}$$

with the root mean square velocity $u_{rms} = \sqrt{\langle \mathbf{u}^2 \rangle}$.

3.2.1.3 The Kolmogorov Microscales

The concept of the energy cascade as discussed in Sect. 3.2 implies the existence of a smallest length scale η at which energy that has been injected at large scales L is ultimately dissipated into heat. According to Kolmogorov's first similarity hypothesis, this characteristic length scale may only depend on the kinematic viscosity ν and the rate at which energy is dissipated, namely, $\langle \varepsilon \rangle$. The latter two possess the dimensions

$$[\nu] = \frac{[m]^2}{[s]} \, , \quad \text{and} \quad [\langle \varepsilon \rangle] = \frac{[m]^2}{[s]^3} \, . \tag{3.2.7}$$

Since $\eta = \eta(\nu, \langle \varepsilon \rangle)$, dimensional analysis suggests that

$$\eta = \left(\frac{\nu^3}{\langle \varepsilon \rangle} \right)^{\frac{1}{4}} \, . \tag{3.2.8}$$

This length scale is usually referred to as Kolmogorov length or Kolmogorov microscale. It is a lower bound for the inertial range, i.e., the cascade process comes to an end in the vicinity of η and the energy is dissipated into heat. The range of scales $r \leq \eta$ is therefore called dissipation range. In the dissipation range, similar considerations can be made for the velocity, time, and wavenumber

$$u_\eta = (\langle \varepsilon \rangle \nu)^{\frac{1}{4}} \, , \quad \tau_\eta = \left(\frac{\nu}{\langle \varepsilon \rangle} \right)^{\frac{1}{2}} \, , \quad \text{and} \quad k_\eta = \left(\frac{\langle \varepsilon \rangle}{\nu^3} \right)^{\frac{1}{4}} \, . \tag{3.2.9}$$

The corresponding Reynolds number of these scales is

$$Re_\eta = \frac{u_\eta \eta}{\nu} = 1 \,. \tag{3.2.10}$$

Therefore, the scales in Eq. (3.2.9) can be considered as characteristic scales of laminar fluid motion.

3.2.1.4 The Taylor Scale

Although the Kolmogorov microscale η is intrinsically defined as dissipative length scale, dissipative effects can already be observed at larger length scales. More precisely, as one follows the turbulent velocity fluctuations from large scales where they possess a certain roughness down to the dissipative scales, one will observe that the velocity field is dominated by smooth solutions. This can be best demonstrated at the example of the auto-correlation function $C(r)/C(0)$ from Eq. (3.2.4) that is schematically depicted in Fig. 3.2: whereas the auto-correlation function is positively curved for large values of r in accordance with the cascade picture, negative curvatures can be measured experimentally for small values of r. Latter feature can be attributed to the influence of dissipation. A characteristic length scale, where the rough character of the velocity field is smoothed out by viscosity can be obtained from a Taylor expansion of the correlation function $C(r)$ around $r = 0$

$$C(r) = C(0) + C'(0)r + \frac{1}{2!}C''(0)r^2 + \text{h.o.t.} \approx C(0)\left(1 - \frac{r^2}{2\lambda^2}\right), \tag{3.2.11}$$

where

$$\lambda = \sqrt{-\frac{C(0)}{C''(0)}} \,. \tag{3.2.12}$$

Here, $C'(0)$ vanishes on the basis of homogeneity which suggests that $C(-r) = \langle u(x-r)u(x)\rangle = \langle u(x)u(x+r)\rangle = C(r)$, where we shifted $x \to x+r$ in the last step. The Taylor microscale λ can be determined by a parabolic fit of the auto-correlation function at the origin, as indicated in Fig. 3.2.

For some purposes, it is convenient to refer to scales other than the integral length scale, for instance, for the definition of Reynolds numbers in experiments that have no comparable boundary conditions. Here, one uses the so-called Taylor-Reynolds number

$$Re_\lambda = \frac{u_{rms}\lambda}{\nu} \,. \tag{3.2.13}$$

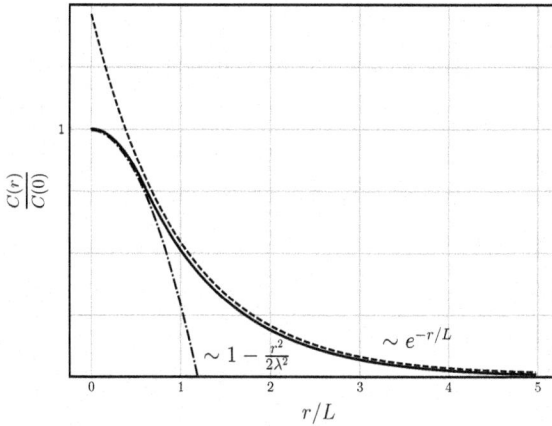

Fig. 3.2 Schematic depiction of the auto-correlation function $C(r)/C(0)$. At large scales, the velocity field at two different points becomes statistically independent, which serves as a definition of the integral length scale L in Eq. (3.2.5). At small scales, the velocity field is smeared out due to the influence of strong viscous forces. These dissipative effects amount in a negative curvature of the auto-correlation function at small r. The Taylor length λ in Eq. (3.2.12) can be obtained from a parabolic fit of the auto-correlation function at the origin

An advantage of the Taylor-Reynolds number is its independence from the integral length scale L, which allows for more reasonable comparisons of experiments with different boundary conditions or energy injection mechanisms.

The integral and Taylor scale limits the so-called inertial range where Kolmogorov's theory applies. Therefore, $\lambda = \lambda(E_{kin}, \nu, \langle \varepsilon \rangle)$ and dimensional analysis suggests that

$$\lambda = \sqrt{\frac{E_{kin}\nu}{\langle \varepsilon \rangle}}. \tag{3.2.14}$$

3.2.2 Kolmogorov's Theory of Locally Isotropic Homogeneous Turbulence

In the following, we adopt Kolmogorov's hypothesis from Sect. 3.2 and deduce important statements for certain statistical quantities. First of all, let us consider the implications of Kolmogorov's second similarity hypothesis: the scale invariance of the Navier-Stokes equation discussed in Sect. 2.1 states that the Navier-Stokes equation is invariant under the transformation

$$\tilde{\mathbf{u}} = \lambda^h \mathbf{u}, \quad \tilde{t} = \lambda^{1-h}t, \quad \tilde{\mathbf{x}} = \lambda \mathbf{x}, \quad \tilde{p} = \lambda^{2h}p, \quad \tilde{\nu} = \lambda^{1+h}\nu. \tag{3.2.15}$$

According to Kolmogorov, the energy dissipation rate ε in Eq. (2.2.1) is a purely local quantity and should be invariant under the transformation (3.2.15), which results in

$$\langle \tilde{\varepsilon}\left(\tilde{\mathbf{x}}\right)\rangle = \frac{\tilde{\nu}}{2} \sum_{i,j} \left\langle \left(\frac{\partial \tilde{u}_i}{\partial \tilde{x}_j} + \frac{\partial \tilde{u}_j}{\partial \tilde{x}_i}\right)^2 \right\rangle = \lambda^{3h-1} \frac{\nu}{2} \sum_{i,j} \left\langle \left(\frac{\partial u_i}{\partial x_j} + \frac{\partial u_j}{\partial x_i}\right)^2 \right\rangle$$

$$= \lambda^{3h-1} \langle \varepsilon(\mathbf{x})\rangle \quad \text{with} \quad h = \frac{1}{3} . \tag{3.2.16}$$

Hence, the scale invariance of ε immediately fixes the exponent $h = 1/3$ and results in an important prediction for the so-called longitudinal structure functions

$$S_n(r) = \langle v(\mathbf{x}, r)^n\rangle = \left\langle \left([\mathbf{u}(\mathbf{x}+\mathbf{r}) - \mathbf{u}(\mathbf{x})] \cdot \frac{\mathbf{r}}{r}\right)^n \right\rangle . \tag{3.2.17}$$

Under the assumption that the nth-order longitudinal structure function follows a power law in the inertial range, i.e., $S_n(r) \sim r^{\zeta_n}$ for $\eta << r << L$, we obtain

$$\underbrace{S_n(r)}_{\sim r^{\zeta_n}} = \left\langle \left([\mathbf{u}(\mathbf{x}+\mathbf{r}) - \mathbf{u}(\mathbf{x})] \cdot \frac{\mathbf{r}}{r}\right)^n \right\rangle$$

$$= \lambda^{-nh} \underbrace{\left\langle \left([\tilde{\mathbf{u}}(\lambda\mathbf{x}+\lambda\mathbf{r}) - \tilde{\mathbf{u}}(\lambda\mathbf{x})] \cdot \frac{\lambda\mathbf{r}}{\lambda r}\right)^n \right\rangle}_{\sim (\lambda r)^{\zeta_n}} . \tag{3.2.18}$$

Using the result $h = 1/3$ from Eq. (3.2.16) in order to cancel λ immediately yields the scaling exponents $\zeta_n = n/3$. According to Kolmogorov's second similarity hypothesis, the inertial range structure function of order n can be deduced from dimensional analysis according to

$$S_n(r) = C_n \langle \varepsilon\rangle^{n/3} r^{n/3} . \tag{3.2.19}$$

Here, the C_n are arbitrary non-dimensional coefficients. As we will see in Sect. 3.3.3, Eq. (3.2.19) is an exact relation for the structure function of third order which characterizes the energy transfer through scales and can be derived directly from the Navier-Stokes equation according to

$$S_3(r) = -\frac{4}{5}\langle \varepsilon\rangle r . \tag{3.2.20}$$

This relation is the famous *Kolmogorov 4/5-law* and implies stationarity, homogeneity, and isotropy of the underlying flow. The extrapolation of this formula to arbitrary order n in Eq. (3.2.19), however, is not guaranteed. In fact, experiments and numerical simulations of the Navier-Stokes equation show deviations from the K41 prediction (3.2.19) that become stronger with increasing order of n. The latter phenomenon of *intermittency* undoubtedly is one of the key signatures of turbulent fluid motion. Before we address this peculiarity in more detail in the following section, it is quite instructive to classify the K41 phenomenology on the basis of probability density functions. Here, it can be shown [6] that the K41 phenomenology implies *self-similar* velocity increment PDFs of the form

$$f(v,r) = \frac{1}{\langle \varepsilon \rangle^{1/3} r^{1/3}} g\left(\frac{v}{\langle \varepsilon \rangle^{1/3} r^{1/3}} \right) . \tag{3.2.21}$$

Taking the moments of the velocity increment, PDF results in the K41 structure functions

$$S_n(r) = \int dv \, v^n f(v,r) = \frac{1}{\langle \varepsilon \rangle^{1/3} r^{1/3}} \int dv \, g\left(\frac{v}{\langle \varepsilon \rangle^{1/3} r^{1/3}} \right)$$

$$= \underbrace{\int dv' v'^n g(v')}_{C_n} \langle \varepsilon \rangle^{n/3} r^{n/3} . \tag{3.2.22}$$

Here, we made a substitution $v' = \frac{v}{\langle \varepsilon \rangle^{1/3} r^{1/3}}$ in the last step. Deviations from the K41 phenomenology for higher orders n can thus be attributed to a *non-self-similar* velocity increment PDF. In the last part of this section, we want to point out another important consequence of the K41 phenomenology: the kinetic energy spectrum $E(k)$, which will be introduced more formally in Sect. 3.3.4.1. The second-order structure function of the K41 phenomenology implies an energy spectrum of the form

$$E(k) = C_K \langle \epsilon \rangle^{2/3} k^{-5/3} . \tag{3.2.23}$$

Here, C_K is the so-called Kolmogorov constant that remains undetermined in Kolmogorov's theory and can be estimated experimentally as $C_K \approx 0.52$ (see also [12] for further discussion). A typical energy spectrum is depicted schematically in Fig. 3.3: Energy is injected into the system at small values of k and is transferred to larger values of k. In the inertial range, the spectrum follows Kolmogorov's prediction $k^{-5/3}$. Adjacent to the inertial range is the dissipation range at large k-values, which is characterized by an exponential decay of the energy spectrum [1].

Fig. 3.3 Schematic
depiction of the energy
spectrum of 3D turbulence:
The energy is injected at the
integral length scale L and
cascades down toward small
scales at the rate $\langle \varepsilon \rangle$ until it is
removed by dissipation

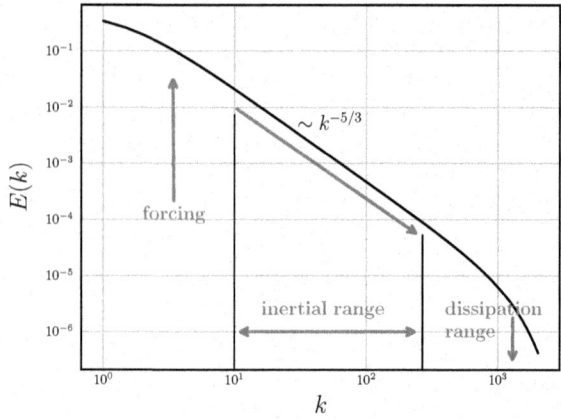

3.2.3   The Phenomenon of Small-Scale Intermittency in Turbulence

As mentioned in the previous section, experiments and numerical simulations indicate deviations from the K41 phenomenology that manifest themselves for increasing
order of the longitudinal structure functions (3.2.17).

Figure 3.4 shows a plot of the scaling exponent ζ_n against order n for the K41
phenomenology next to experimental results and various other phenomenologies that
were devised to overcome the difficulties of the K41 phenomenology. The $n/3$-line
(K41) corresponds well to experimentally determined values for small n. For $n = 3$,
we have the exact result (3.2.20) which is satisfied by all phenomenologies as well
as by the experiment.

At larger values of n, however, the experimental values indicate a nonlinear order
dependence of ζ_n that is in contradiction to the K41 predictions. Latter phenomenon
is referred to as intermittency (it is not to be confused with the notion of intermittency
in chaotic systems, such as the logistic map). It can also be observed on the basis
of probability density functions (PDFs) of longitudinal velocity increments $f(v, r)$,
where deviations from the self-similar K41 form (3.2.21) are present at small scales
and manifest themselves in form of an *increased probability for large negative and
large positive longitudinal velocity increment fluctuations*. Figure 3.5 shows the
longitudinal velocity increment PDF $f(v, r)$ at different scales obtained from a free
jet experiment [16]. Here, the large-scale PDF is close to a Gaussian distribution,
since the two velocity field fluctuations $\mathbf{u}(\mathbf{x} + \mathbf{r})$ and $\mathbf{u}(\mathbf{x})$ that enter the increment
become statistically independent for large $\mathbf{r}$ (the one-point velocity field PDF is close
to Gaussian). However, at smaller scales, the PDFs show pronounced tails which are
characteristics for a turbulent velocity field.

The cause for the phenomenon of small-scale intermittency in turbulent flows is
still not fully understood. As Landau already pointed out shortly after Kolmogorov's
phenomenology was published [17], the K41 theory solely contains the mean local

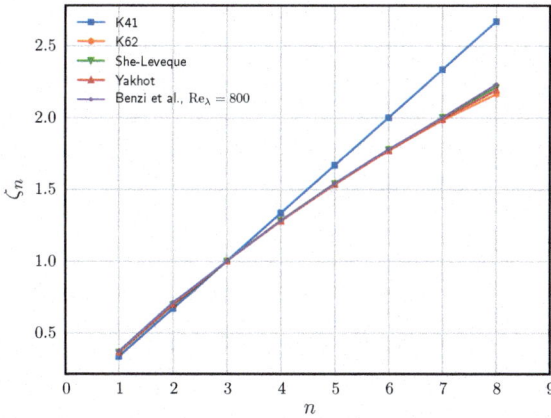

Fig. 3.4 Scaling exponent ζ_n of longitudinal structure functions against n. The K41 phenomenology behaves as $n/3$, whereas experiments from Benzi et al. [13] indicate a nonlinear order dependence of the scaling exponent. Furthermore, the plot shows scaling exponents that belong to the K62 phenomenology, the She-Leveque phenomenology, as well as to a phenomenology derived by Yakhot [14]

Fig. 3.5 Probability density function of the longitudinal velocity increments $f(v, r)$ for different scales r from experiments by Renner [15]. The PDFs are shifted vertically for better clarity

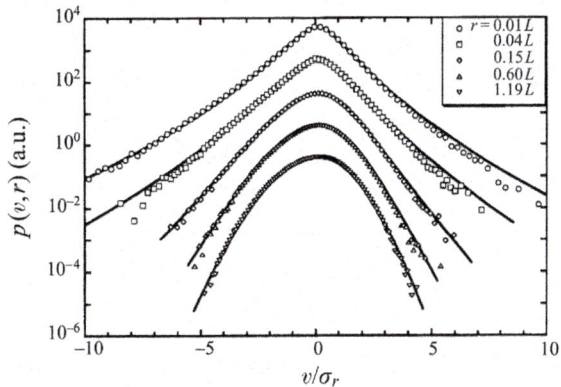

energy dissipation rate $\langle \varepsilon \rangle$. Latter is fixed by the assumption of constant energy transfer through scales. This hypothesis, however, entirely neglects a random character of the local energy dissipation rate ε. Since dominant geometrical structures in turbulent flows apparently consist of vortex tubes and sheets that lead to a nearly singular character of the velocity field in their vicinity (see also Sect. 2.3.2), the local energy dissipation rate must strongly fluctuate from point to point. This hypothesis, which is at odds with the K41 theory, is also strongly supported by numerical simulations [6].

3.2.3.1   Kolmogorov's Refined Similarity Hypothesis

In contemplation of adjusting these flaws of the K41 theory, Kolmogorov himself [18] and Oboukhov [19] extended formula (3.2.21) in order to allow for fluctuations of the local energy dissipation rate. In proposing a log-normal distribution for the fluctuations of ε (see [6] for further details), Kolmogorov was able to derive scaling exponents of the form

$$\zeta_n = \frac{n}{3} - \frac{\mu}{18} n(n-3) , \tag{3.2.24}$$

with an undetermined intermittency coefficient μ. As it can readily be seen, the so-called K62 scaling fulfills the requirement $\zeta_3 = 1$. Moreover, experiments suggest a value of $\mu \approx 0.227$ for the intermittency coefficient [20]. The K62 model obeys a nonlinear order dependence as depicted in Fig. 3.4. Hence, it conforms better with experimentally determined scaling exponents than the K41 phenomenology. Nevertheless, a major shortcoming of the K62 model is that for increasing order n, more precisely at $n \geq 3/2 + 3/\mu$, the scaling exponents decrease due to the parabolic form of Eq. (3.2.24). This finding is in contradiction with a convexity condition for structure functions [20], which states $\zeta_{2n+2} \geq \zeta_{2n}$.

3.2.3.2   The Model by She and Leveque

Another intermittency model that attunes well to numerical and experimental results for structure function exponents ζ_n is the model by She and Leveque [21]. Latter fixes the scaling behavior for $n \to \infty$ under the assumption that singular structures of a turbulent flow consist of flux tubes similar to the ones discussed in Sect. 2.3.2. Further restrains on the corresponding values of ζ_n at $n = 3$ ($\zeta_3 = 1$) and $n = 0$ ($\zeta_0 = 0$) imply scaling exponents

$$\zeta_n = \frac{n}{9} + 2 \left(1 - \left(\frac{2}{3} \right)^{n/3} \right) . \tag{3.2.25}$$

The K62 and She-Leveque scaling belong to the family of so-called multifractal scaling which will be further described in Sect. 5.1.2.

3.3   Moment Formulation: The Friedmann-Keller Hierarchy

So far, we have only focused on phenomenological models that bear little to no relation to the deterministic equations of turbulent fluid motion. Consequently, the next sections are concerned with the determination of evolution equations for certain statistical quantities, such as correlation functions or probability density functions, derived directly from the Navier-Stokes equation. In the following, we want to focus on the moment formulation of turbulence, which originates from the work of O. Reynolds who focused on one-point statistics of turbulent flows [22]. The pioneering work by O. Reynolds in 1883 was continued by G. I. Taylor who introduced the notion of correlation functions of random velocity fields [23] to turbulence. Similar concepts were used in 1924 by the two Russian mathematicians A. A. Friedmann and L. V. Keller. Later on, Friedmann became known for his theory of the expanding universe on the basis of Einstein's equations (we refer the reader to the excellent biographical monologue [24]). In their at that time unapprehended work [25], Friedmann and Keller attempted to derive a closed equation for the rate of change of the two-point correlation function $C_{ij}(\mathbf{x}, \mathbf{x}', t) = \langle u_i(\mathbf{x}, t) u_j(\mathbf{x}', t) \rangle$. Latter can be derived directly from the Navier-Stokes equation (2.1.1) according to

$$
\frac{\partial}{\partial t} \langle u_i(\mathbf{x}, t) u_j(\mathbf{x}', t) \rangle
$$

$$
+ \sum_k \frac{\partial}{\partial x_k} \langle u_k(\mathbf{x}, t) u_i(\mathbf{x}, t) u_j(\mathbf{x}', t) \rangle + \sum_k \frac{\partial}{\partial x_k'} \langle u_k(\mathbf{x}', t) u_i(\mathbf{x}, t) u_j(\mathbf{x}', t) \rangle
$$

$$
= - \frac{\partial}{\partial x_i} \langle p(\mathbf{x}, t) u_j(\mathbf{x}', t) \rangle - \frac{\partial}{\partial x_j'} \langle p(\mathbf{x}', t) u_i(\mathbf{x}, t) \rangle
$$

$$
+ \nu [\Delta_{\mathbf{x}} + \Delta_{\mathbf{x}'}] \langle u_i(\mathbf{x}, t) u_j(\mathbf{x}', t) \rangle + Q_{ij}(\mathbf{x}, \mathbf{x}', t) . \tag{3.3.1}
$$

The important observation by Friedmann and Keller was that Eq. (3.3.1) is part of an *infinite* hierarchy of transport equations. Thereby, due to the nonlinear and nonlocal character of the underlying Navier-Stokes equation, the moment equation of order n depends on the moment of order $n + 1$ leaving us with an infinite hierarchy of moment equations, the Friedmann- Keller hierarchy. At this point, it is important to mention that this so-called closure problem of turbulence originates not only from the convective part, but mainly arises due to nonlocal pressure contributions in the moment equations. Despite the fact that the system of transport equations gives an analytical formulation of the problem of turbulence, every subsystem is necessarily non-closed, i.e., it contains more unknowns than equations in the given subsystem.

3.3.1 Reynolds-Averaged Navier-Stokes Equation

For most practical purposes, e.g., in engineering, it is not necessary to know every detail of the fluctuating fields. To this end, we assume that the velocity field can be divided into mean and fluctuating parts according to

$$\mathbf{u}(\mathbf{x}, t) = \bar{\mathbf{u}}(\mathbf{x}, t) + \mathbf{u}'(\mathbf{x}, t) \, , \qquad (3.3.2)$$

$$p(\mathbf{x}, t) = \bar{p}(\mathbf{x}, t) + p'(\mathbf{x}, t) \, , \qquad (3.3.3)$$

where the temporal mean values of the fluctuating parts vanish

$$\langle \mathbf{u}'(\mathbf{x}, t) \rangle = \lim_{T \to \infty} \frac{1}{T} \int_t^{t+T} dt' \, \mathbf{u}'(\mathbf{x}, t') = 0 \, . \qquad (3.3.4)$$

Furthermore, terms like $\langle \mathbf{u}'(\mathbf{x}, t) \cdot \nabla \mathbf{u}(\mathbf{x}, t) \rangle$ vanish, since the mean streaming profile is assumed to remain constant in the interval T. The Reynolds-averaged Navier-Stokes equation reads

$$\frac{D}{Dt} \bar{\mathbf{u}}(\mathbf{x}, t) + \bar{\mathbf{u}}(\mathbf{x}, t) \cdot \nabla \bar{\mathbf{u}}(\mathbf{x}, t) + \nabla \cdot \langle \mathbf{u}'(\mathbf{x}, t) \mathbf{u}'(\mathbf{x}, t) \rangle = -\nabla \bar{p}(\mathbf{x}, t) + \nu \Delta \bar{\mathbf{u}}(\mathbf{x}, t) \, ,$$
$$(3.3.5)$$

where we introduced the total derivative $\frac{D}{Dt} = \frac{\partial}{\partial t} + \bar{u}_n \frac{\partial}{\partial x_n}$ with summation over the same index. This equation can be written in component form according to

$$\frac{D}{Dt} \bar{u}_i = \frac{\partial}{\partial x_j} \left[\nu \left(\frac{\partial \bar{u}_i}{\partial x_j} + \frac{\partial \bar{u}_j}{\partial x_i} \right) - \bar{p} \delta_{ij} - \langle u'_i u'_j \rangle \right] , \qquad (3.3.6)$$

which corresponds to the Navier-Stokes equation of the mean velocity field with an additional nonlinearity. Latter term represents the turbulent pulsations which have a non-negligible influence on the mean fields. The tensor which enters Eq. (3.3.6),

$$R_{ij} = \langle u'_i u'_j \rangle \, , \qquad (3.3.7)$$

is known as Reynolds' stress tensor. Knowledge of this tensor in terms of the mean field would signify a considerable step forward in the calculations of mean streaming profiles as, for example, the flow behind a wedge. So far no such knowledge has come to existence, and therefore one has to model turbulent stresses by an effective damping term that characterizes energy flux from the mean field to the small-scale turbulent fine structure.

In ordinary turbulence, Reynolds stresses are written in analogous form to the viscous stress tensor [26] used in the derivation of the Navier-Stokes equation

$$R_{ij} = \langle u'_i u'_j \rangle = -\nu_t \left(\frac{\partial \bar{u}_i}{\partial x_j} + \frac{\partial \bar{u}_j}{\partial x_i} \right) + \frac{1}{3} \langle u'_k u'_k \rangle \delta_{ij} \, . \qquad (3.3.8)$$

The so-called eddy viscosity ν_t is thereby much larger than the kinematic viscosity ν and in the simplest model one assumes that it depends only on the turbulent kinetic energy $K = \frac{1}{2} \langle u'_i u'_i \rangle$ and the local energy dissipation rate ε,

$$\nu_t = \frac{C K^2}{\langle \varepsilon \rangle} \, , \qquad (3.3.9)$$

where C is a free numerical parameter.

Another interesting model goes back to L. Prandtl [27] and is based on the mixing length r_m,

$$\nu_t = |\mathbf{u}'|r_m \ . \tag{3.3.10}$$

The mixing length is a turbulent "mean free path" and an additional transport equation for the turbulent kinetic energy has to be solved.

3.3.2 The von Karman-Howarth Equation

As it has been mentioned in the introduction of this chapter, the concept of the turbulent energy cascade is detached from the Navier-Stokes equation. Therefore today, one of the main objectives in turbulence theory is to obtain information on the transport of energy through scales directly from the Navier-Stokes equation. In the following, we want to simplify the transport equation (3.3.1) under the assumption of homogeneity, isotropy, and mirror symmetry.

- Since we are assuming homogeneity, correlation functions solely depend on the relative distance $\mathbf{r} = \mathbf{x}' - \mathbf{x}$, so that

$$C_{ij}(\mathbf{r}, t) = \langle u_i(\mathbf{x}, t)u_j(\mathbf{x}', t)\rangle = C_{ij}(-\mathbf{r}, t) \ , \tag{3.3.11}$$

$$C_{ki,j}(\mathbf{r}, t) = \langle u_k(\mathbf{x}, t)u_i(\mathbf{x}, t)u_j(\mathbf{x}', t)\rangle \ , \tag{3.3.12}$$

$$C_{ki,j}(-\mathbf{r}, t) = \langle u_k(\mathbf{x}', t)u_i(\mathbf{x}', t)u_j(\mathbf{x}, t)\rangle \ . \tag{3.3.13}$$

Therefore, viscous terms can be rewritten according to

$$[\Delta_{\mathbf{x}} + \Delta_{\mathbf{x}'}]\langle u_i(\mathbf{x}, t)u_j(\mathbf{x}', t)\rangle = 2\Delta_{\mathbf{r}}\langle u_i(\mathbf{x}, t)u_j(\mathbf{x}', t)\rangle = 2\Delta_{\mathbf{r}}C_{ij}(\mathbf{r}, t) \ . \tag{3.3.14}$$

Moreover, isotropic and mirror symmetric tensors of third order fulfill the following relation:

$$C_{kj,i}(-\mathbf{r}, t) = -C_{ki,j}(\mathbf{r}, t) \ . \tag{3.3.15}$$

- Pressure-velocity correlations vanish on the basis of isotropy and mirror symmetry [6] which yields

$$\langle p(\mathbf{x}, t)u_j(\mathbf{x}', t)\rangle = 0 \ . \tag{3.3.16}$$

It should be noted that this is only true at this stage of the hierarchy, relating two-point and three-point quantities. Higher order transport equations contain non-vanishing pressure-velocity correlations that represent a significant difficulty for closure approximations of the Friedmann-Keller hierarchy.

Under the above assumptions Eq. (3.3.1) reduces to

$$\frac{\partial}{\partial t}C_{ij}(\mathbf{r}, t) - 2\frac{\partial}{\partial r_k}C_{ki,j}(\mathbf{r}, t) = 2\nu\Delta_{\mathbf{r}}C_{ij}(\mathbf{r}, t) , \qquad (3.3.17)$$

where summation over equal indices is implied. A further simplification of Eq. (3.3.17) can be obtained from the invariant theory of isotropic and homogenous turbulence. Each tensor can solely be rewritten in terms of its longitudinal correlation function, which is a consequence of the incompressibility condition and is further derived in Appendix 3. One obtains

$$C_{ij}(\mathbf{r}, t) = \left(C_{rr}(r, t) - \frac{1}{2r}\frac{\partial}{\partial r}(r^2 C_{rr}(r, t)) \right)\frac{r_i r_j}{r^2} + \frac{1}{2r}\frac{\partial}{\partial r}(r^2 C_{rr}(r, t))\delta_{ij} ,$$
$$(3.3.18)$$

and

$$C_{ki,j}(\mathbf{r}, t) = -\frac{r^2}{2}\frac{\partial}{\partial r}\left(\frac{C_{rrr}(r, t)}{r}\right)\frac{r_i r_j r_k}{r^3}$$
$$+ \frac{1}{4r}\frac{\partial}{\partial r}\left(r^2 C_{rrr}(r, t)\right)\left(\frac{r_i}{r}\delta_{kj} + \frac{r_k}{r}\delta_{ij}\right) - \frac{C_{rrr}(r, t)}{2}\frac{r_j}{r}\delta_{ik} .$$
$$(3.3.19)$$

Let us sum over equal indices i and j in Eqs. (3.3.18–3.3.19). We define

$$Q_{kin}(r, t) = \sum_{i=j} C_{ij}(\mathbf{r}, t) , \qquad (3.3.20)$$

$$J_k^{kin}(\mathbf{r}, t) = -\sum_{i=j} C_{ki,j}(\mathbf{r}, t) , \qquad (3.3.21)$$

and obtain a balance equation for $Q_{kin}(r, t)$ with a corresponding current $\mathbf{J}^{kin}(\mathbf{r}, t)$ that reads

$$\frac{\partial}{\partial t}Q_{kin}(r, t) + 2\frac{\partial}{\partial r_k}J_k^{kin}(\mathbf{r}, t) = 2\nu\Delta_{\mathbf{r}}Q_{kin}(r, t) . \qquad (3.3.22)$$

$Q_{kin}(r, t)$ and its corresponding current $\mathbf{J}^{kin}(\mathbf{r}, t)$ can be expressed in terms of $C_{rr}(r, t)$, $C_{rrr}(r, t)$ and $C_{ttr}(r, t)$ as it is shown in Appendix 3. In exchanging the derivatives with respect to t and r, we obtain

$$\frac{\partial}{\partial t}Q_{kin}(r, t) = \frac{1}{r^2}\frac{\partial}{\partial r}\left(r^3\frac{\partial}{\partial t}C_{rr}(r, t), \right), \qquad (3.3.23)$$

$$2\frac{\partial}{\partial r_k}J_k^{kin}(\mathbf{r}, t) = -\frac{1}{r^2}\frac{\partial}{\partial r}\left(\frac{1}{r}\frac{\partial}{\partial r}\left[r^4 C_{rrr}(r, t)\right]\right) , \qquad (3.3.24)$$

$$\Delta_\mathbf{r} Q_{kin}(r, t) = \frac{1}{r^2} \frac{\partial}{\partial r} \left(r^2 \frac{\partial}{\partial r} Q_{kin}(r, t) \right) . \tag{3.3.25}$$

Inserting these relations into Eq. (3.3.22) finally yields the so-called von Kármán-Howarth equation of fluid turbulence [28]

$$\frac{\partial}{\partial t} C_{rr}(r, t) = \frac{1}{r^4} \frac{\partial}{\partial r} \left[r^4 \left(C_{rrr}(r, t) + 2\nu \frac{\partial}{\partial r} C_{rr}(r, t) \right) \right] . \tag{3.3.26}$$

Although this equation still contains two unknowns in the form of the longitudinal correlation functions of second and third orders, and thus appears to be of no particular further use, it should be noted that the von Kármán-Howarth equation leads to a series of important consequences: First of all, it serves as a starting point for the derivation of the famous 4/5-law in turbulence as we will show in the next section. Moreover, it can be used to derive an evolution equation for longitudinal vorticity correlations, and it also makes certain assertions concerning the final period of decaying turbulence, where the nonlinearity in Eq. (3.3.26) becomes negligible.

3.3.3 Kolmogorov's 4/5-Law

As we already mentioned in the section discussing Kolmogorov's theory of local isotropic turbulence 3.2.2, the Navier-Stokes equation exhibits an exact relation for the third-order structure function. This relation, the so-called *4/5-law*, is considered to be one of the few exact relations derived directly from the Navier-Stokes equation. Here, we will give a rather cursory derivation from the von Kármán-Howarth equation (3.3.26). First, we will replace the third-order correlation function by the longitudinal structure function of third order $C_{rrr}(r, t) = S_{rrr}(r, t)/6$. Second, we will subtract the ensemble-averaged equation (2.2.2) and recover an evolution equation for the longitudinal structure function of second order $S_{rr}(r, t) = 2(C_{rr}(0, t) - C_{rr}(r, t))$ that reads

$$\frac{1}{2} \frac{\partial}{\partial t} S_{rr}(r, t) = -\frac{2}{3} \langle \varepsilon \rangle - \frac{1}{r^4} \frac{\partial}{\partial r} \left[r^4 \left(\frac{1}{6} S_{rrr}(r, t) + 2\nu \frac{\partial}{\partial r} S_{rr}(r, t) \right) \right] . \tag{3.3.27}$$

Under the assumption of stationarity, we can neglect the temporal evolution of $S_{rr}(r, t)$, multiply by r^4, and integrate with respect to r, which yields

$$S_{rrr}(r, t) = -\frac{4}{5} \langle \varepsilon \rangle r + \frac{\nu}{6} \frac{\partial}{\partial r} S_{rr}(r, t) . \tag{3.3.28}$$

In the inertial range, i.e., for $\eta \ll r \ll L$, the last viscous term can be neglected and we obtain Kolmogorov's 4/5-law, $S_{rrr}(r, t) = -\frac{4}{5}\langle\varepsilon\rangle r$. Latter suggests a nonlinear transfer of energy from large to small scales whereby the transfer rate is given by $\langle\varepsilon\rangle$. It necessarily implies a non-Gaussian longitudinal velocity increment PDF, due to a non-vanishing moment of third order $S_{rrr}(r, t) \neq 0$ in the inertial range.

In the dissipation range, for $r \leq \eta$, the third-order structure function behaves as $S_{rrr} \sim r^3$, which can be seen from a Taylor expansion. Therefore, it has no contribution in the dissipation range (the nonlinear energy transfer is conserved) and we obtain

$$0 = -\frac{4}{5}\langle\varepsilon\rangle r + \frac{\nu}{6}\frac{\partial}{\partial r}S_{rr}(r, t) , \qquad (3.3.29)$$

which yields a smooth second-order structure function in the dissipation range, $S_{rr}(r, t) = \langle\varepsilon\rangle r^2/15\nu$.

3.3.4   The Moment Hierarchy in Fourier Space

The Navier-Stokes equation in Fourier space can be obtained by introducing the Fourier transform of the velocity field $\mathbf{u}(\mathbf{x}, t)$ according to

$$\hat{u}_i(\mathbf{k}, t) = \frac{1}{(2\pi)^3} \int d\mathbf{x} e^{-i\mathbf{k}\cdot\mathbf{x}} u_i(\mathbf{x}, t) . \qquad (3.3.30)$$

A Fourier transform of the Navier-Stokes equation yields

$$\left(\frac{\partial}{\partial t} + \nu k^2\right)\hat{u}_i(\mathbf{k}, t) = -ik_l \int d\mathbf{k}' \hat{u}_i(\mathbf{k} - \mathbf{k}', t)\hat{u}_l(\mathbf{k}', t) - ik_i \hat{p}(\mathbf{k}, t) , \quad (3.3.31)$$

where we implied summation over equal indices. Again, pressure contributions can be determined in making use of the incompressibility condition, i.e., in taking the dot product of Eq. (3.3.31) with $\mathbf{k}$

$$k_i k_l \int d\mathbf{k}' \hat{u}_i(\mathbf{k} - \mathbf{k}', t)\hat{u}_l(\mathbf{k}', t) + k^2 \hat{p}(\mathbf{k}, t) = 0 . \qquad (3.3.32)$$

Consequently, the Navier-Stokes equation in Fourier space reads

$$\left(\frac{\partial}{\partial t} + \nu k^2\right)\hat{u}_i(\mathbf{k}, t) = -ik_l \left(\delta_{ij} - \frac{k_i k_j}{k^2}\right) \int d\mathbf{k}' \hat{u}_j(\mathbf{k} - \mathbf{k}', t)\hat{u}_l(\mathbf{k}', t) .$$

$$(3.3.33)$$

In order to derive the spectral version of the moment hierarchy (3.3.1) from the Navier-Stokes equation in Fourier space (3.3.33) it is appropriate to investigate how the assumption of homogeneity translates into Fourier space. To this end, we consider velocity correlations between two wave vectors $\mathbf{k}$ and $\mathbf{k}'$

$$\langle \hat{u}_i(\mathbf{k}, t)\hat{u}_j(\mathbf{k}', t)\rangle = \frac{1}{(2\pi)^6} \int d\mathbf{x} \int d\mathbf{x}' e^{-i\mathbf{k}\cdot\mathbf{x} - i\mathbf{k}'\cdot\mathbf{x}'} \langle u_i(\mathbf{x}, t)u_j(\mathbf{x}, t)\rangle . \qquad (3.3.34)$$

Setting $\mathbf{x}' = \mathbf{x} + \mathbf{r}$ and making use of Eq. (3.3.11) yield

$$\langle \hat{u}_i(\mathbf{k}, t)\hat{u}_j(\mathbf{k}', t)\rangle = \frac{1}{(2\pi)^6} \int d\mathbf{x} \int d\mathbf{r} e^{-i(\mathbf{k}+\mathbf{k}')\cdot\mathbf{x} - i\mathbf{k}'\cdot\mathbf{r}} C_{ij}(\mathbf{r}, t) \qquad (3.3.35)$$

$$= \frac{1}{(2\pi)^3} \delta(\mathbf{k} + \mathbf{k}') \int d\mathbf{r} e^{-i\mathbf{k}'\cdot\mathbf{r}} C_{ij}(\mathbf{r}, t) = \delta(\mathbf{k} + \mathbf{k}')\hat{C}_{ij}(\mathbf{k}, t) .$$

Therefore, non-vanishing two-point correlations in Fourier space only occur for velocity fluctuations with antiparallel wave vectors $\mathbf{k} = -\mathbf{k}'$. Moreover, it can be shown [29] that all multi-wavenumber moments vanish unless their wave vectors add to zero, e.g.,

$$\langle \hat{u}_i(\mathbf{k}, t)\hat{u}_j(\mathbf{k}', t)\hat{u}_l(\mathbf{k}'', t)\rangle = 0 , \qquad \text{unless} \quad \mathbf{k} + \mathbf{k}' + \mathbf{k}'' = 0 . \qquad (3.3.36)$$

We are now in the position to formulate the moment hierarchy in Fourier space starting from the two wavenumber tensor $\hat{C}_{ij}(\mathbf{k}, t)$ from Eq. (3.3.36). To this end, we multiply Eq. (3.3.33) by $\hat{u}_j(\mathbf{k}, t)$ and repeat this procedure for interchanged indices. Adding these two equations together and taking the ensemble average yield

$$\left(\frac{\partial}{\partial t} + \nu k^2 + \nu k'^2\right) \underbrace{\langle \hat{u}_i(\mathbf{k}, t)\hat{u}_j(\mathbf{k}', t)\rangle}_{=\hat{C}_{ij}(\mathbf{k},t)\delta(\mathbf{k}+\mathbf{k}')}$$

$$= -ik_l\left(\delta_{im} - \frac{k_i k_m}{k^2}\right) \underbrace{\int d\mathbf{k}'' \langle \hat{u}_m(\mathbf{k} - \mathbf{k}'', t)\hat{u}_l(\mathbf{k}'', t)\hat{u}_j(\mathbf{k}', t)\rangle}_{=\delta(\mathbf{k}+\mathbf{k}')\hat{C}_{ml,j}(-\mathbf{k},t)}$$

$$-ik_l'\left(\delta_{jm} - \frac{k_j' k_m'}{k'^2}\right) \underbrace{\int d\mathbf{k}'' \langle \hat{u}_m(\mathbf{k}' - \mathbf{k}'', t)\hat{u}_l(\mathbf{k}'', t)\hat{u}_i(\mathbf{k}, t)\rangle}_{=\delta(\mathbf{k}+\mathbf{k}')\hat{C}_{ml,i}(\mathbf{k},t)} , \qquad (3.3.37)$$

where $\hat{C}_{ml,j}(\mathbf{k}, t)$ is the Fourier transform of $C_{ml,j}(\mathbf{r}, t)$ in Eq. (3.3.12). We thus obtain

$$\left(\frac{\partial}{\partial t} + 2\nu k^2\right) \hat{C}_{ij}(\mathbf{k}, t)$$

$$= -ik_l\left(\delta_{im} - \frac{k_i k_m}{k^2}\right) \hat{C}_{ml,j}(-\mathbf{k}, t) + ik_l\left(\delta_{jm} - \frac{k_j k_m}{k^2}\right) \hat{C}_{ml,i}(\mathbf{k}, t) . \qquad (3.3.38)$$

A further simplification of these equations is achieved in making use of Eq. (3.3.15) which yields the balance equation for the spectral tensor of second order $\hat{C}_{ij}(\mathbf{k}, t)$ according to

$$\left(\frac{\partial}{\partial t} + 2\nu k^2\right) \hat{C}_{ij}(\mathbf{k}, t) + 2ik_l\hat{C}_{li,j}(\mathbf{k}, t) = 0 . \tag{3.3.39}$$

Of course, this result could have been obtained directly by Fourier transforming Eq. (3.3.17). For later convenience, we specify an evolution equation that determines the unknown current in Eq. (3.3.39), namely,

$$\left(\frac{\partial}{\partial t} + \nu k^2 + \nu k'^2 + \nu|\mathbf{k} - \mathbf{k}'|^2\right) \langle \hat{u}_i(-\mathbf{k}, t)\hat{u}_j(\mathbf{k}', t)\hat{u}_l(\mathbf{k} - \mathbf{k}', t)\rangle$$

$$= +ik_m\left(\delta_{in} - \frac{k_ik_n}{k^2}\right)$$

$$\times \int d\mathbf{k}'' \langle \hat{u}_n(-\mathbf{k} - \mathbf{k}'', t)\hat{u}_m(\mathbf{k}'', t)\hat{u}_j(\mathbf{k}', t)\hat{u}_l(\mathbf{k} - \mathbf{k}', t)\rangle$$

$$-ik'_m\left(\delta_{jn} - \frac{k'_jk'_n}{k^2}\right)$$

$$\times \int d\mathbf{k}'' \langle \hat{u}_n(\mathbf{k}' - \mathbf{k}'', t)\hat{u}_m(\mathbf{k}'', t)\hat{u}_i(-\mathbf{k} - \mathbf{k}', t)\hat{u}_l(\mathbf{k} - \mathbf{k}', t)\rangle$$

$$- i(k_m - k'_m)\left(\delta_{ln} - \frac{(k_l - k'_l)(k_n - k'_n)}{|\mathbf{k} - \mathbf{k}'|^2}\right)$$

$$\times \int d\mathbf{k}'' \langle \hat{u}_n(\mathbf{k} - \mathbf{k}' - \mathbf{k}'', t)\hat{u}_m(\mathbf{k}'', t)\hat{u}_i(-\mathbf{k}, t)\hat{u}_j(\mathbf{k}', t)\rangle . \tag{3.3.40}$$

Again, due to nonlinear contributions we are left with an infinite hierarchy of transport equations. Regarding closure schemes for the latter, it is often more convenient to work in Fourier space, since viscous contributions become local. Here, the starting point for many closure assumptions is the evolution equation for the energy spectrum, which will be introduced in the next section.

3.3.4.1   The Evolution Equation for the Energy Spectrum

A further elucidation of the energy transfer in homogeneous and isotropic turbulence can be gained from the evolution equation of the energy spectrum $E(k, t)$ that is the determining scalar of the spectral tensor of second order

$$\hat{C}_{ij}(\mathbf{k}, t) = \frac{1}{4\pi k^2}\left(-\frac{k_ik_j}{k^2} + \delta_{ij}\right) E(k, t) . \tag{3.3.41}$$

Summing Eq. (3.3.39) over equal indices i and j yields the evolution equation for the energy spectrum

$$\left(\frac{\partial}{\partial t} + 2\nu k^2\right) E(k,t) + T(k,t) = Q(k,t) \,. \tag{3.3.42}$$

Here, $Q(k,t)$ denotes an additional source term stemming from external forcing and

$$T(k,t) = 4\pi i k^2 \sum_{i=j} k_l \hat{C}_{li,j}(\mathbf{k},t) \,, \tag{3.3.43}$$

represents an unknown flux term that responsible for energy transfer between different wavenumbers. In fact, we obtain [1]

$$\int_0^\infty dk \, T(k,t) = 0 \,. \tag{3.3.44}$$

Hence, the transfer term reorganizes energy among different k-values suggesting that variations in total kinetic energy are solely caused by viscous forces and external forcing. The equation for the energy spectrum Eq. (3.3.42) is the point of departure for numerous closure approximations, which involve certain assumptions for the unknown energy transfer term (3.3.43) discussed in Sect. 4.1.

3.4 Kinetic Equations of Turbulent Fluid Motion

Another approach to a statistical description of turbulence is based on kinetic equations of turbulent motion. Latter bear similarities with the BBGKY hierarchy from statistical mechanics which will be demonstrated via the example of multi-point velocity PDFs. The kinetic approach emanated from ideas of Lundgren [30] and Monin [31], who considered multi-point velocity PDFs, and Novikov [32], who focused on multi-point vorticity PDFs. For an overview of kinetic equations in turbulence, we also refer the reader to the review article by Friedrich et al. [33].

3.4.1 The Lundgren-Monin-Novikov Hierarchy for Multi-Point Velocity PDFs in Turbulence

Here, we follow the seminal paper by Lundgren [30] who derived evolution equations for the multi-point velocity distribution functions. To this end, we start from the so-called fine-grained one-point velocity PDF

$$\hat{f}_1(\mathbf{u}_1, \mathbf{x}_1, t) = \delta(\mathbf{u}_1 - \mathbf{u}(\mathbf{x}_1, t)) \,, \tag{3.4.1}$$

which represents a sharp distribution in the space of the sample variable $\mathbf{u}_1$ for a given realization of the velocity field $\mathbf{u}(\mathbf{x}_1, t)$. In order to obtain the one-point PDF $f_1(\mathbf{u}_1, \mathbf{x}_1, t)$ from this quantity, we have to perform an ensemble average over all possible realizations of the velocity field according to

$$f_1(\mathbf{u}_1, \mathbf{x}_1, t) = \langle \hat{f}_1(\mathbf{u}_1, \mathbf{x}_1, t) \rangle = \langle \delta(\mathbf{u}_1 - \mathbf{u}(\mathbf{x}_1, t)) \rangle \ . \tag{3.4.2}$$

In the same manner, we can define multi-point PDFs like the two-point PDF

$$f_2(\mathbf{u}_2, \mathbf{x}_2; \mathbf{u}_1, \mathbf{x}_1, t) = \langle \delta(\mathbf{u}_2 - \mathbf{u}(\mathbf{x}_2, t)) \delta(\mathbf{u}_1 - \mathbf{u}(\mathbf{x}_1, t)) \rangle \ , \tag{3.4.3}$$

or the multi-point PDF

$$f_n(\mathbf{u}_n, \mathbf{x}_n; \ldots; \mathbf{u}_1, \mathbf{x}_1, t) = \prod_{i=1}^{n} \langle \delta(\mathbf{u}_i - \mathbf{u}(\mathbf{x}_i, t)) \rangle \ . \tag{3.4.4}$$

The n-point PDF is convenient mean to describe spatial correlations at a given instance in time.[1] At this point, it is worth mentioning certain beneficial properties of the multi-point PDF. First of all, the reduction property of the n-point PDFs,

$$\int \mathrm{d}\mathbf{u}_n \, f_n(\mathbf{u}_n, \mathbf{x}_n; \mathbf{u}_{n-1}, \mathbf{x}_{n-1}; \ldots; \mathbf{u}_1, \mathbf{x}_1, t) = f_{n-1}(\mathbf{u}_{n-1}, \mathbf{x}_{n-1}; \ldots; \mathbf{u}_1, \mathbf{x}_1, t) \ ,$$
$$\tag{3.4.5}$$

follows immediately from definition (3.4.4) and reduces n-point statistics to $(n-1)$-point statistics. Moreover, the coincidence property states that the fusion of two points reduces the n-point PDF to

$$\int \mathrm{d}\mathbf{x}_n \, \delta(\mathbf{x}_n - \mathbf{x}_{n-1}) f_n(\mathbf{u}_n, \mathbf{x}_n; \mathbf{u}_{n-1}, \mathbf{x}_{n-1}; \ldots; \mathbf{u}_1, \mathbf{x}_1, t)$$
$$= \delta(\mathbf{u}_n - \mathbf{u}_{n-1}) f_{n-1}(\mathbf{u}_{n-1}, \mathbf{x}_{n-1}; \ldots; \mathbf{u}_1, \mathbf{x}_1, t) \ , \tag{3.4.6}$$

meaning that the velocities $\mathbf{u}_n$ and $\mathbf{u}_{n-1}$ are fully correlated. The opposite case is the statistical independence, e.g., for $|\mathbf{x}_2 - \mathbf{x}_1| \to \infty$, the two-point PDF factorizes to

$$f_2(\mathbf{u}_2, \mathbf{x}_2; \mathbf{u}_1, \mathbf{x}_1, t) = f_1(\mathbf{u}_1, \mathbf{x}_1, t) f_1(\mathbf{u}_2, \mathbf{x}_2, t) \ . \tag{3.4.7}$$

In the following, we want to relate these statistical quantities to the basic fluid dynamical equations, which in this case is the Navier-Stokes equation (2.1.1) and the incompressibility condition for the velocity field (2.1.2). Therefore, the procedure consists of deriving the PDF (3.4.2) with respect to time

---

[1]Clearly, joint PDFs for different times and different points can be defined in the same fashion but are of no particular interest here and we refer the reader to [1] for further discussion.

$$\frac{\partial}{\partial t} f_1(\mathbf{u}_1, \mathbf{x}_1, t) = -\left\langle \underbrace{\frac{\partial \mathbf{u}(\mathbf{x}_1, t)}{\partial t}}_{\text{Eq.(2.1.1)}} \cdot \nabla_{\mathbf{u}_1} \hat{f}_1(\mathbf{u}_1, \mathbf{x}_1, t) \right\rangle \tag{3.4.8}$$

$$= \left\langle \left[\mathbf{u}(\mathbf{x}_1, t) \cdot \nabla_{\mathbf{x}_1} \mathbf{u}(\mathbf{x}_1, t) + \nabla_{\mathbf{x}_1} p(\mathbf{x}_1, t) - \nu \Delta_{\mathbf{x}_1} \mathbf{u}(\mathbf{x}_1, t) \right] \cdot \nabla_{\mathbf{u}_1} \hat{f}_1(\mathbf{u}_1, \mathbf{x}_1, t) \right\rangle,$$

where the chain rule was used in the first step. In the following, each term is treated separately. Starting with the advective term we obtain

$$\left\langle \left[\mathbf{u}(\mathbf{x}_1, t) \cdot \nabla_{\mathbf{x}_1} \mathbf{u}(\mathbf{x}_1, t) \right] \cdot \nabla_{\mathbf{u}_1} \delta(\mathbf{u}_1 - \mathbf{u}(\mathbf{x}_1, t)) \right\rangle$$

$$= -\left\langle \mathbf{u}(\mathbf{x}_1, t) \cdot \nabla_{\mathbf{x}_1} \delta(\mathbf{u}_1 - \mathbf{u}(\mathbf{x}_1, t)) \right\rangle \underbrace{=}_{\text{Eq.(2.1.2)}} -\nabla_{\mathbf{x}_1} \cdot \left\langle \mathbf{u}(\mathbf{x}_1, t) \delta(\mathbf{u}_1 - \mathbf{u}(\mathbf{x}_1, t)) \right\rangle$$

$$= -\nabla_{\mathbf{x}_1} \cdot \left\langle \mathbf{u}_1 \delta(\mathbf{u}_1 - \mathbf{u}(\mathbf{x}_1, t)) \right\rangle = -\mathbf{u}_1 \cdot \nabla_{\mathbf{x}_1} f_1(\mathbf{u}_1, \mathbf{x}_1, t). \tag{3.4.9}$$

In the first step, we used the reverse chain rule and then made use of the so-called sifting property of the delta function, $\mathbf{u}(\mathbf{x}_1, t)\delta(\mathbf{u}_1 - \mathbf{u}(\mathbf{x}_1, t)) = \mathbf{u}_1\delta(\mathbf{u}_1 - \mathbf{u}(\mathbf{x}_1, t))$. This enabled us to pull the sample variable $\mathbf{u}_1$ out of the ensemble average in the last step. Turning to pressure contributions, we have to make use of the pressure representation in Eq. (2.1.8) and obtain

$$\left\langle \left[\nabla_{\mathbf{x}_1} p(\mathbf{x}_1, t) \right] \cdot \nabla_{\mathbf{u}_1} \delta(\mathbf{u}_1 - \mathbf{u}(\mathbf{x}_1, t)) \right\rangle$$

$$= \nabla_{\mathbf{u}_1} \cdot \left\langle \left(\int d\mathbf{x}_2 \left(\nabla_{\mathbf{x}_1} \frac{\nabla_{\mathbf{x}_2} \cdot (\mathbf{u}(\mathbf{x}_2, t) \cdot \nabla_{\mathbf{x}_2} \mathbf{u}(\mathbf{x}_2, t))}{4\pi |\mathbf{x}_1 - \mathbf{x}_2|} \right) \delta(\mathbf{u}_1 - \mathbf{u}(\mathbf{x}_1, t)) \right\rangle$$

$$= \nabla_{\mathbf{u}_1} \cdot \int d\mathbf{x}_2 \left(\nabla_{\mathbf{x}_1} \frac{1}{4\pi |\mathbf{x}_1 - \mathbf{x}_2|} \right)$$

$$\frac{\partial}{\partial x_{2,i}} \frac{\partial}{\partial x_{2,j}} \left\langle \underbrace{\left(u_j(\mathbf{x}_2, t) u_i(\mathbf{x}_2, t) \right)}_{\int d\mathbf{u}_2 u_{2,j} u_{2,i} \delta(\mathbf{u}_2 - \mathbf{u}(\mathbf{x}_2, t))} \delta(\mathbf{u}_1 - \mathbf{u}(\mathbf{x}_1, t)) \right\rangle$$

$$= \nabla_{\mathbf{u}_1} \cdot \int d\mathbf{x}_2 \left(\nabla_{\mathbf{x}_1} \frac{1}{4\pi |\mathbf{x}_1 - \mathbf{x}_2|} \right) \int d\mathbf{u}_2 \left(\mathbf{u}_2 \cdot \nabla_{\mathbf{x}_2} \right)^2$$

$$\underbrace{\left\langle \delta(\mathbf{u}_2 - \mathbf{u}(\mathbf{x}_2, t)) \delta(\mathbf{u}_1 - \mathbf{u}(\mathbf{x}_1, t)) \right\rangle}_{= f_2(\mathbf{u}_2, \mathbf{x}_2; \mathbf{u}_1, \mathbf{x}_1, t)}. \tag{3.4.10}$$

Due to the nonlocality of the pressure, this contribution couples to the two-point velocity PDF $f_2(\mathbf{u}_2, \mathbf{x}_2; \mathbf{u}_1, \mathbf{x}_1, t)$ which must be considered as an unknown and has to be determined from the next order equation. This is also true for the viscous term since it involves second-order velocity field derivatives which cannot be related to the one-increment PDF. Accordingly, the viscous term is also unclosed which yields

$$\nu \left\langle [\Delta_{\mathbf{x}_1} (\mathbf{u}(\mathbf{x}_1, t)] \cdot \nabla_{\mathbf{u}_1} \delta(\mathbf{u}_1 - \mathbf{u}(\mathbf{x}_1, t)) \right\rangle \qquad (3.4.11)$$

$$= \nu \nabla_{\mathbf{u}_1} \cdot \int d\mathbf{x}_2 \delta(\mathbf{x}_2 - \mathbf{x}_1) \Delta_{\mathbf{x}_2} \left\langle \mathbf{u}(\mathbf{x}_2, t) \delta(\mathbf{u}_1 - \mathbf{u}(\mathbf{x}_1, t)) \right\rangle$$

$$= \nu \nabla_{\mathbf{u}_1} \cdot \int d\mathbf{x}_2 \delta(\mathbf{x}_2 - \mathbf{x}_1) \Delta_{\mathbf{x}_2} \int d\mathbf{u}_2 \mathbf{u}_2 \underbrace{\left\langle \delta(\mathbf{u}_2 - \mathbf{u}(\mathbf{x}_2, t)) \delta(\mathbf{u}_1 - \mathbf{u}(\mathbf{x}_1, t)) \right\rangle}_{= f_2(\mathbf{u}_2, \mathbf{x}_2; \mathbf{u}_1, \mathbf{x}_1, t)} .$$

Inserting Eqs. (3.4.9), (3.4.10), and (3.4.11) into Eq. (3.4.8) yields

$$\frac{\partial}{\partial t} f_1(\mathbf{u}_1, \mathbf{x}_1, t) + \mathbf{u}_1 \cdot \nabla_{\mathbf{x}_1} f_1(\mathbf{u}_1, \mathbf{x}_1, t)$$

$$= \nabla_{\mathbf{u}_1} \cdot \int d\mathbf{x}_2 \left(\nabla_{\mathbf{x}_1} \frac{1}{4\pi |\mathbf{x}_1 - \mathbf{x}_2|} \right) \int d\mathbf{u}_2 \left(\mathbf{u}_2 \cdot \nabla_{\mathbf{x}_2} \right)^2 f_2(\mathbf{u}_2, \mathbf{x}_2; \mathbf{u}_1, \mathbf{x}_1, t)$$

$$- \nu \nabla_{\mathbf{u}_1} \cdot \int d\mathbf{x}_2 \delta(\mathbf{x}_2 - \mathbf{x}_1) \Delta_{\mathbf{x}_2} \int d\mathbf{u}_2 \mathbf{u}_2 f_2(\mathbf{u}_2, \mathbf{x}_2; \mathbf{u}_1, \mathbf{x}_1, t) . \qquad (3.4.12)$$

This is the first equation in an infinite series of multi-point velocity PDF equations. In fact, one can derive an evolution equation for the n-point PDF (3.4.4) which, due to the product rule, reveals an additive structure of the multi-point PDF hierarchy [34]. The evolution equation of the n-point PDF reads

$$\frac{\partial}{\partial t} f_n(\{\mathbf{u}_i, \mathbf{x}_i\}, t) + \sum_{i=1}^{n} \mathbf{u}_i \cdot \nabla_{\mathbf{x}_i} f_n(\{\mathbf{u}_i, \mathbf{x}_i\}, t)$$

$$= \sum_{i=1}^{n} \nabla_{\mathbf{u}_i} \cdot \int d\mathbf{x} \left(\nabla_{\mathbf{x}_i} \frac{1}{4\pi |\mathbf{x}_i - \mathbf{x}|} \right) \int d\mathbf{u} \left(\mathbf{u} \cdot \nabla_{\mathbf{x}} \right)^2 f_{n+1}(\mathbf{u}, \mathbf{x}; \{\mathbf{u}_i, \mathbf{x}_i\}, t)$$

$$- \nu \sum_{i=1}^{n} \nabla_{\mathbf{u}_i} \cdot \int d\mathbf{x} \delta(\mathbf{x} - \mathbf{x}_i) \Delta_{\mathbf{x}} \int d\mathbf{u} \mathbf{u} f_{n+1}(\mathbf{u}, \mathbf{x}; \{\mathbf{u}_i, \mathbf{x}_i\}, t) . \qquad (3.4.13)$$

In contrast to the Friedmann-Keller hierarchy that leads to unclosed nonlinear terms, the Lundgren-Monin-Novikov hierarchy (3.4.13) possesses *closed advective* terms, i.e., the nonlinearity in the Navier-Stokes equation can be traced back to a Liouville-type expression of the n-point PDF. The hierarchical character of Eq. (3.4.13), hence, arises on the basis of pressure and viscous terms, which necessarily involve the $(n + 1)$-point PDF. Furthermore, we must emphasize that the Lundgren-Monin-Novikov hierarchy is able to reproduce the entire Friedmann-Keller hierarchy through taking the moments of the n-point PDF from Eq. (3.4.13).

3.5 The Functional Formulation of the Problem of Turbulence

So far we discussed the moment formulation (Friedmann-Keller hierarchy) and the PDF formulation (Lundgren-Monin-Novikov hierarchy) of turbulence. A further approach toward a complete statistical description of the velocity field is the so-called functional formalism of turbulence [1, 35]. The central quantity of this approach is the characteristic functional

$$\varphi[\alpha(\mathbf{x}, t)] = \left\langle e^{i \int \mathrm{d}\mathbf{x}\mathrm{d}t\, \alpha(\mathbf{x},t)\cdot\mathbf{u}(\mathbf{x},t)} \right\rangle . \tag{3.5.1}$$

We can already see that values at the characteristic points,

$$\alpha(\mathbf{x}, t) = \sum_{k=1}^{n} \alpha_k \delta(\mathbf{x} - \mathbf{x}_k)\delta(t - t_k) , \tag{3.5.2}$$

coincide with the ordinary characteristic function, i.e., the Fourier transform of the n-time-point PDF

$$\varphi(\alpha_1, \mathbf{x}_1, t_1; \alpha_2, \mathbf{x}_2, t_2; \ldots, \alpha_n, \mathbf{x}_n, t_n) = \left\langle e^{i \sum_{k=1}^{n} \alpha_k \cdot \mathbf{u}(\mathbf{x}_k, t_k)} \right\rangle . \tag{3.5.3}$$

Therefore, the characteristic functional is an appropriate mean for the determination of all finite-dimensional PDFs. A less complete statistical description involves the spatial characteristic functional

$$\varphi[\alpha(\mathbf{x}), t] = \left\langle e^{i \int \mathrm{d}\mathbf{x}\, \alpha(\mathbf{x})\cdot\mathbf{u}(\mathbf{x},t)} \right\rangle , \tag{3.5.4}$$

which is sufficient for a complete statistical description at a given time t but contains no information on joint characteristics at different times. An important restriction for the characteristic functional results from the incompressibility condition for the velocity field (2.1.2). Latter can be cast in the form

$$\int_V \mathrm{d}\mathbf{x}\, \mathbf{u}(\mathbf{x}) \cdot \nabla\psi(\mathbf{x}) \underbrace{=}_{\nabla\cdot\mathbf{u}=0} \int_V \mathrm{d}\mathbf{x}\, \nabla \cdot (\mathbf{u}(\mathbf{x})\psi(\mathbf{x})) = \oint_S \mathrm{d}\mathbf{a} \cdot \mathbf{u}(\mathbf{x})\psi(\mathbf{x}) = 0 , \tag{3.5.5}$$

where $\psi(\mathbf{x})$ is an arbitrary scalar function and where we assumed that the velocity field vanishes at the fluid boundaries. We thus find

$$(\alpha(\mathbf{x}) + \nabla\psi(\mathbf{x})) \cdot \mathbf{u}(\mathbf{x}) = \alpha(\mathbf{x}) \cdot \mathbf{u}(\mathbf{x}) + \underbrace{\nabla\psi(\mathbf{x}) \cdot \mathbf{u}(\mathbf{x})}_{=0,\ \mathrm{Eq.}(3.5.5)} . \tag{3.5.6}$$

Hence, the characteristic functional is invariant under the transformation $\alpha_k(\mathbf{x}) \to \alpha_k(\mathbf{x}) + \frac{\partial}{\partial x_k}\psi(\mathbf{x})$. Furthermore, it can be shown [1] that it is sufficient to specify this restriction only for the initial condition $\varphi[\alpha(\mathbf{x}), t = 0] = \varphi_0[\alpha(\mathbf{x})]$ which yields

$$\varphi_0[\alpha(\mathbf{x})] = \varphi_0[\alpha(\mathbf{x}) + \nabla\psi(\mathbf{x})] \ . \tag{3.5.7}$$

In order to obtain the temporal evolution of the characteristic functional (3.5.4), we have to draw on the calculus of functional derivatives. A first useful relation derived from the spatial characteristic functional reads

$$\frac{\delta\varphi[\alpha(\mathbf{x}), t]}{\delta\alpha_j(\mathbf{x}'')} = i \int d\mathbf{x}' \left\langle \underbrace{\frac{\delta(\alpha_k(\mathbf{x}')u_k(\mathbf{x}', t))}{\delta\alpha_j(\mathbf{x}'')}}_{=\delta_{jk}\delta(\mathbf{x}'-\mathbf{x}'')u_k(\mathbf{x}',t)} e^{i\int d\mathbf{x}\alpha(\mathbf{x})\cdot\mathbf{u}(\mathbf{x},t)} \right\rangle$$

$$= i \left\langle u_j(\mathbf{x}'', t)e^{i\int d\mathbf{x}\alpha(\mathbf{x})\cdot\mathbf{u}(\mathbf{x},t)} \right\rangle \ , \tag{3.5.8}$$

where summation over equal indices is implied. The temporal evolution of the spatial characteristic functional reads

$$\frac{\partial\varphi[\alpha(\mathbf{x}), t]}{\partial t} = i \int d\mathbf{x}'\alpha_k(\mathbf{x}') \left\langle \underbrace{\frac{\partial u_k(\mathbf{x}', t)}{\partial t}}_{\text{Eq.(2.1.1)}} e^{i\int d\mathbf{x}\alpha(\mathbf{x})\cdot\mathbf{u}(\mathbf{x},t)} \right\rangle$$

$$= i \int d\mathbf{x}'\alpha_k(\mathbf{x}') \left\langle \left[-\frac{\partial}{\partial x_j'}u_j(\mathbf{x}', t)u_k(\mathbf{x}', t) - \frac{\partial}{\partial x_k'}p(\mathbf{x}', t) + \nu\Delta_{\mathbf{x}'}u_k(\mathbf{x}', t) \right] \right.$$

$$\left. \times e^{i\int d\mathbf{x}\alpha(\mathbf{x})\cdot\mathbf{u}(\mathbf{x},t)} \right\rangle \ , \tag{3.5.9}$$

where we applied the incompressibility of the velocity field in the nonlinear term. The terms are evaluated with the help of our useful relation (3.5.8) and only the pressure term has to be rewritten as

$$\left\langle p(\mathbf{x}', t)e^{i\int d\mathbf{x}\alpha(\mathbf{x})\cdot\mathbf{u}(\mathbf{x},t)} \right\rangle$$

$$= -\frac{1}{4\pi} \int \frac{d\mathbf{x}''}{|\mathbf{x}' - \mathbf{x}''|} \left\langle \frac{\partial^2(u_l(\mathbf{x}'', t)u_j(\mathbf{x}'', t))}{\partial x_l''\partial x_j''} e^{i\int d\mathbf{x}\alpha(\mathbf{x})\cdot\mathbf{u}(\mathbf{x},t)} \right\rangle$$

$$= -\frac{1}{4\pi} \int \frac{d\mathbf{x}''}{|\mathbf{x}' - \mathbf{x}''|} \frac{\partial^2}{\partial x_l''\partial x_j''} \frac{\delta^2}{\delta\alpha_l(\mathbf{x}'')\delta\alpha_j(\mathbf{x}'')} \underbrace{\left\langle e^{i\int d\mathbf{x}\alpha(\mathbf{x})\cdot\mathbf{u}(\mathbf{x},t)} \right\rangle}_{\varphi[\alpha(\mathbf{x}),t]} \ . \tag{3.5.10}$$

Therefore, the temporal evolution equation for the spatial characteristic functional in closed form reads

$$\frac{\partial \varphi[\boldsymbol{\alpha}(\mathbf{x}), t]}{\partial t} \tag{3.5.11}$$

$$= \int \mathrm{d}\mathbf{x}' \alpha_k(\mathbf{x}') \left[i \frac{\partial}{\partial x_j'} \frac{\delta^2}{\delta \alpha_j(\mathbf{x}')\delta \alpha_k(\mathbf{x}')} \right.$$

$$\left. + \frac{\partial}{\partial x_k'} \int \frac{\mathrm{d}\mathbf{x}''}{4\pi|\mathbf{x}' - \mathbf{x}''|} \frac{\partial^2}{\partial x_l'' \partial x_j''} \frac{\delta^2}{\delta \alpha_l(\mathbf{x}'')\delta \alpha_j(\mathbf{x}'')} + \nu \Delta_{\mathbf{x}'} \frac{\delta}{\delta \alpha_k(\mathbf{x}')} \right] \varphi[\boldsymbol{\alpha}(\mathbf{x}), t] \ .$$

Moreover, the pressure term can be eliminated with the help of the incompressibility condition, i.e., setting $\alpha_k(\mathbf{x}) = \tilde{\alpha}_k(\mathbf{x}) + \frac{\partial}{\partial x_k}\psi(\mathbf{x})$ which yields

$$\frac{\partial \varphi[\boldsymbol{\alpha}(\mathbf{x}), t]}{\partial t} \tag{3.5.12}$$

$$= \int \mathrm{d}\mathbf{x}' \tilde{\alpha}_k(\mathbf{x}') \left\{ i \frac{\partial}{\partial x_j'} \frac{\delta^2}{\delta \alpha_j(\mathbf{x}')\delta \alpha_k(\mathbf{x}')} + \nu \Delta_{\mathbf{x}'} \frac{\delta}{\delta \alpha_k(\mathbf{x}')} \right\} \varphi[\boldsymbol{\alpha}(\mathbf{x}), t].$$

Equation (3.5.12) or the equivalent Eq. (3.5.13) was first derived by Hopf [35] and is referred to as Hopf equation [1]. It is considered as the most compact formulation of statistical hydrodynamics. Moreover, since Eq. (3.5.12) is a linear equation for the spatial characteristic functional $\varphi[\boldsymbol{\alpha}(\mathbf{x}), t]$ it entails that the statistical dynamics of a turbulent flow is a linear problem, although its individual realizations are described by nonlinear partial differential equations. The linearity of the Hopf equation necessarily implies that the spatial characteristic functional follows the superposition principle: if the characteristic functionals $\varphi^{(k)}[\boldsymbol{\alpha}(\mathbf{x}), t]$ are solutions of Eq. (3.5.12) for the corresponding initial conditions $\varphi^{(k)}[\boldsymbol{\alpha}(\mathbf{x}), t = 0] = \varphi_0^{(k)}[\boldsymbol{\alpha}(\mathbf{x})]$, then the functional $\varphi[\boldsymbol{\alpha}(\mathbf{x}), t]$ is a linear combination of these solutions. Although these rather surprising properties of the Hopf equation seem to be quite promising, we must emphasize that solving Eq. (3.5.12) proves to be challenging from a mathematical point of view. It is commonly believed that new mathematical tools for solving such linear functional equations have yet to be found. Moreover, existence and uniqueness theorems for solutions of Eq. (3.5.12) are still an unresolved matter in the mathematical branch of turbulence theory. Nevertheless, the functional formulation has certain advantages which include its compatibility with moment and kinetic formulations. For the case of moments, this can best be seen in expanding the characteristic functional (3.5.4) in a power series according to

$$\varphi[\alpha(\mathbf{x}), t] = \varphi^{(0)} + \varphi^{(1)} + \varphi^{(2)} + \dots \tag{3.5.13}$$

$$= \varphi[0, t] + \int d\mathbf{x}' \left. \frac{\delta\varphi[\alpha(\mathbf{x}), t]}{\delta\alpha_k(\mathbf{x}')} \right|_{\alpha(\mathbf{x})=0} \alpha_k(\mathbf{x}')$$

$$+ \frac{1}{2!} \int d\mathbf{x}'' \int d\mathbf{x}' \left. \frac{\delta^2\varphi[\alpha(\mathbf{x}), t]}{\delta\alpha_j(\mathbf{x}'')\delta\alpha_k(\mathbf{x}')} \right|_{\alpha(\mathbf{x})=0} \alpha_j(\mathbf{x}'')\alpha_k(\mathbf{x}') + \text{h.o.t.}$$

$$= 1 + i \int d\mathbf{x} \langle u_k(\mathbf{x}, t) \rangle \alpha_k(\mathbf{x}) + \frac{i^2}{2!} \int d\mathbf{x}' \int d\mathbf{x} \langle u_j(\mathbf{x}', t)u_k(\mathbf{x}, t) \rangle \alpha_j(\mathbf{x}')\alpha_k(\mathbf{x})$$

$$+ \text{h.o.t.}$$

$$= 1 + \sum_{n=1}^{\infty} \int \dots \int d\mathbf{x}_1 \dots d\mathbf{x}_n \frac{i^n}{n!} \alpha_{k_1}(\mathbf{x}_1) \dots \alpha_{k_n}(\mathbf{x}_n) C_{k_1 \dots k_n}(\mathbf{x}_1, \dots, \mathbf{x}_n, t) .$$

Hence, the knowledge of *all* n-point moments,

$$C_{k_1 \dots k_n}(\mathbf{x}_1, \dots, \mathbf{x}_n, t) = \langle u_{k_1}(\mathbf{x}_1, t)u_{k_2}(\mathbf{x}_2, t) \dots u_{k_n}(\mathbf{x}_n, t) \rangle , \tag{3.5.14}$$

is necessary for the determination of the characteristic functional. By the same token, moments can be derived from the characteristic functional according to

$$C_{k_1 \dots k_n}(\mathbf{x}_1, \dots, \mathbf{x}_n, t) = \frac{1}{i^n} \left. \frac{\delta^n\varphi[\alpha(\mathbf{x}), t]}{\delta\alpha_{k_1}(\mathbf{x}_1) \dots \delta\alpha_{k_n}(\mathbf{x}')} \right|_{\alpha(\mathbf{x})=0} . \tag{3.5.15}$$

Inserting the power series (3.5.13) into the Hopf equation (3.5.13) and matching terms of equal order yield

$$\frac{\partial\varphi^{(n)}[\alpha(\mathbf{x}), t]}{\partial t} \tag{3.5.16}$$

$$= \int d\mathbf{x}' \tilde{\alpha}_k(\mathbf{x}') \left\{ i \frac{\partial}{\partial x'_j} \left(\frac{\delta^2\varphi^{(n+1)}[\alpha(\mathbf{x}), t]}{\delta\alpha_j(\mathbf{x}')\delta\alpha_k(\mathbf{x}')} \right) + \nu\Delta_{\mathbf{x}'} \left(\frac{\delta\varphi^{(n)}[\alpha(\mathbf{x}), t]}{\delta\alpha_k(\mathbf{x}'')} \right) \right\} .$$

Again, this represents an infinite series of differential equations for correlation functions (3.5.15) and is thus equivalent to the Friedmann-Keller hierarchy (3.3.1). For some purposes, it is convenient to define the cumulants $K_{k_1, \dots k_n}(\mathbf{x}_1, \dots, \mathbf{x}_n, t)$ of the characteristic functional according to

$$\varphi[\alpha(\mathbf{x}), t] \tag{3.5.17}$$

$$= 1 + \sum_{n=1}^{\infty} \int d\mathbf{x}_1 \dots \int d\mathbf{x}_n \frac{i^n}{n!} \alpha_{k_1}(\mathbf{x}_1) \dots \alpha_{k_n}(\mathbf{x}_n) C_{k_1 \dots k_n}(\mathbf{x}_1, \dots, \mathbf{x}_n, t)$$

$$= \exp\left(\sum_{n=1}^{\infty} \int d\mathbf{x}_1 \dots \int d\mathbf{x}_n \frac{i^n}{n!} \alpha_{k_1}(\mathbf{x}_1) \dots \alpha_{k_n}(\mathbf{x}_n) K_{k_1 \dots k_n}(\mathbf{x}_1, \dots, \mathbf{x}_n, t) \right) .$$

This relation suggests that the first n cumulants can be expressed by the first n moments according to

$$K_i = C_i , \tag{3.5.18}$$

$$K_{ij} = C_{ij} - C_i C_j , \tag{3.5.19}$$

$$K_{ijk} = C_{ijk} - C_{ij} C_k - C_{ik} C_j - C_{jk} C_i + 2 C_i C_j C_k , \tag{3.5.20}$$

$$K_{ijkl} = C_{ijkl} - C_{ij} C_{kl} - C_{ik} C_{jl} - C_{il} C_{jk} \tag{3.5.21}$$
$$- C_{ijk} C_l - C_{ijl} C_k - C_{ikl} C_j - C_{jkl} C_i$$
$$+ 2 C_{ij} C_k C_l + 2 C_{ik} C_j C_l + 2 C_{il} C_k C_l$$
$$+ 2 C_{kl} C_i C_j + 2 C_{jk} C_i C_l + 2 C_{jl} C_i C_k - 6 C_i C_j C_k C_l .$$

The relations simplify in the absence of a mean flow which implies the vanishing of all moments of first order C_i. It is sometimes useful to consider the implications of vanishing cumulants. For instance, the case where all cumulants with the exception of the first two vanish yields

$$\varphi[\alpha(\mathbf{x}), t] = \exp \left[-\frac{1}{2} \int \int d\mathbf{x}'' d\mathbf{x}' \alpha_j(\mathbf{x}'') K_{jk}(\mathbf{x}'', \mathbf{x}', t) \alpha_k(\mathbf{x}') \right] . \tag{3.5.22}$$

This approximation corresponds to a Gaussian approximation of the multi-point velocity PDFs and has been used as a closure method for the Lundgren-Monin-Novikov hierarchy of the one-point velocity PDF equation [36, 37].

3.6 Chapter Conclusions

The current chapter was concerned with statistical formulations of turbulence. We discussed the concept of turbulent energy transfer, i.e., a transport process of kinetic energy from large to small scales via a hierarchy of instabilities of turbulent eddies. Here, the influence of turbulent eddies on the energy transfer was presented in a rather disheveled way in Sect. 2.3.2. Obviously, the turbulent energy cascade in the sense of Kolmogorov and Richardson runs into severe difficulties that are mostly related to small-scale effects of intermittency which dictate strongly non-self-similar behavior of turbulent velocity field fluctuations.

The observation of intermittency in experiments and numerical simulations, the log-normal (K62) as well as the She-Leveque phenomenology were briefly discussed in Sect. 3.2.3. It was stressed that, although the She-Leveque model agrees well with experimental observations, it is detached from the basic hydrodynamic equations. The main part of this chapter dealt with the so-called closure problem of turbulence: the derivation of evolution equations for velocity field moments (correlation functions) led to an infinite hierarchy of moment evolution equations, the Friedmann-Keller hierarchy (3.3.1). Here, unclosed terms were due to the nonlinearity of the

Navier-Stokes equation. Furthermore, a hierarchy of evolution equations for the multi-point velocity field PDFs was derived in Sect. 3.4. This so-called Lundgren-Monin-Novikov hierarchy exhibits unclosed terms due to pressure and viscous contributions. Finally, we described a functional formulation of turbulence via the Hopf functional of turbulence. The evolution equation for the characteristic functional has the remarkable property that it is linear and closed. However, it must be stressed that a solution of the Hopf equation proves to be challenging from a mathematical point of view. Accordingly, the following chapter gives an overview of attempts that try to cope directly with the closure problem in turbulence.

Appendix 1: The Calculus of Isotropic Tensors

For a comprehensive overview of the calculus of isotropic tensors, we refer the reader to [6–8]. In the following, we focus on homogeneous and isotropic tensor fields. Considering the second-order tensor

$$A_{ij}(\mathbf{x}, \mathbf{x} + \mathbf{r}, t) = \langle u_i(\mathbf{x}, t) u_j(\mathbf{x} + \mathbf{r}, t) \rangle , \tag{3.6.1}$$

we conclude that under the assumption of homogeneity, $A_{ij}(\mathbf{x}, \mathbf{x} + \mathbf{r}, t)$ depends solely on the relative distance $\mathbf{r}$ between the velocity field $u_i(\mathbf{x}, t)$ and $u_j(\mathbf{x} + \mathbf{r}, t)$, and not on $\mathbf{x}$

$$A_{ij}(\mathbf{x}, \mathbf{x} + \mathbf{r}, t) = A_{ij}(\mathbf{r}, t) . \tag{3.6.2}$$

Turning next to the symmetry of isotropy, the basic idea is that the tensor $A_{ij}(\mathbf{r}, t)$ has to be invariant under rotations. This means that it has to be invariant under a coordinate transform

$$\mathbf{r}' = U\mathbf{r} \qquad U^t U = 1 , \tag{3.6.3}$$

where $U \in$ SO(3).

This implies that

$$A'_{ij}(\mathbf{r}', t) = \sum_{kl} U_{ik} U_{jl} A_{ij}(\mathbf{r}', t) \tag{3.6.4}$$

has to be equal to $A_{ij}(\mathbf{r}, t)$, which yields

$$A_{ij}(\mathbf{r}, t) = \sum_{kl} U_{ik} U_{jl} A_{ij}(U^t \mathbf{r}, t) . \tag{3.6.5}$$

One can show that for this tensor of second order only the basic tensors $r_i r_j / r^2$, δ_{ij} and $\varepsilon_{ijk} r_k / r$ fulfill the invariance condition. A general tensor of second order can thus be written as

$$A_{ij}(\mathbf{r}, t) = A_1(r, t)\frac{r_i r_j}{r^2} + A_2(r, t)\delta_{ij} + A_3(r, t)\varepsilon_{ijk}\frac{r_k}{r} . \qquad (3.6.6)$$

For the next order, we retrieve

$$A_{ijk}(\mathbf{r}, t) = A_1(r, t)\frac{r_i r_j r_k}{r^3} + A_2(r, t)\delta_{ij}\frac{r_k}{r}$$
$$+ A_3(r, t)\delta_{ik}\frac{r_j}{r} + A_4(r, t)\delta_{jk}\frac{r_i}{r} + A_5(r, t)\varepsilon_{ijk} . \qquad (3.6.7)$$

The last basic tensor involving the ε-tensor can be omitted if $A_{ijk}(\mathbf{r}, t)$ is invariant under reflections, which is certainly the case for the tensor considered in (3.6.6), since both u_i and u_j change their sign under a reflection. Nevertheless, tensors like $B_{ijk}(\mathbf{r}, t) = \langle \omega_i(\mathbf{x} + \mathbf{r}, t)u_j(\mathbf{x}, t)u_k(\mathbf{x}, t)\rangle$ are not invariant under these reflections since only u_j and u_k change their sign, whereas ω_i is an axial vector that keeps its orientation [7]. The resulting tensor is skew and can be written as

$$B_{ijk}(\mathbf{r}, t) = B_1(r, t)\varepsilon_{ijk} . \qquad (3.6.8)$$

Appendix 2: Longitudinal and Transverse Correlation Functions

If we consider the velocity fields $\mathbf{u}(\mathbf{x} + \mathbf{r}, t)$ at point $\mathbf{x} + \mathbf{r}$ and $\mathbf{u}(\mathbf{x}, t)$ at point $\mathbf{x}$, we can divide the vector $\mathbf{u}$ into a part $\mathbf{u}^l$ parallel to $\mathbf{r}$ and a transverse part $\mathbf{u}^t$. These parts are thereby given as

$$\mathbf{u}^l = \frac{\mathbf{r}}{r}\left(\frac{\mathbf{r}}{r}\cdot\mathbf{u}\right), \qquad (3.6.9)$$

$$\mathbf{u}^t = -\left(\frac{\mathbf{r}}{r}\times\left(\frac{\mathbf{r}}{r}\times\mathbf{u}\right)\right) . \qquad (3.6.10)$$

The longitudinal correlation function

$$C_{rr}(r, t) = \langle \mathbf{u}^l(\mathbf{x}, t)\cdot\mathbf{u}^l(\mathbf{x} + \mathbf{r}, t)\rangle \qquad (3.6.11)$$

can be calculated in multiplying the two-point correlation tensor $C_{ij}(\mathbf{r}, t) = \langle u_i(\mathbf{x}, t)u_j(\mathbf{x} + \mathbf{r}, t)\rangle$ by r_i and r_j.

Assuming that $C_{ij}(\mathbf{r}, t) = \langle u_i(\mathbf{x}, t)u_j(\mathbf{x} + \mathbf{r}, t)\rangle$ is isotropic and mirror symmetric, it follows from (3.6.6) that its general form is given as

$$C_{ij}(\mathbf{r}, t) = (C_{rr}(r, t) - C_{tt}(r, t))\frac{r_i r_j}{r^2} + C_{tt}(r, t)\delta_{ij}, \qquad (3.6.12)$$

where $C_{tt}(r, t)$ is the transverse correlation function

$$C_{tt}(r, t) = \langle \mathbf{u}^t(\mathbf{x} + \mathbf{r}, t) \cdot \mathbf{u}^t(\mathbf{x}, t) \rangle . \tag{3.6.13}$$

Turning next to the third-order correlation function

$$C_{i\,j,k}(\mathbf{r}, t) = \langle u_i(\mathbf{x}, t) u_j(\mathbf{x}, t) u_k(\mathbf{x} + \mathbf{r}, t) \rangle , \tag{3.6.14}$$

we conclude that it is symmetric in i and j, and therefore $A_3(r, t) = A_4(r, t)$ in (3.6.7). Furthermore, we obtain the longitudinal correlation function in multiplying Eq. (3.6.7) by $r_i r_j r_k$

$$C_{rrr}(r, t) = A_1(r, t) + A_2(r, t) + 2A_3(r, t) . \tag{3.6.15}$$

Appendix 3: The Correlation Functions for Incompressible, Isotropic, and Homogeneous Fields

The Correlation Function of Second Order

Due to the incompressibility condition it is possible to reduce the tensorial form of $C_{i\,j}(\mathbf{r}, t)$ to a dependence of the longitudinal structure function $C_{rr}(r, t)$ only. The incompressibility condition is therefore used according to

$$\frac{\partial}{\partial r_i} C_{i\,j}(\mathbf{r}, t) = \left\langle \frac{\partial u_i(\mathbf{x} + \mathbf{r}, t)}{\partial r_i} u_j(\mathbf{x}, t), \right\rangle = 0 , \tag{3.6.16}$$

where the summation over equal indices is implied.

By making use of the relation $\frac{\partial}{\partial r_i} = \frac{r_i}{r} \frac{\partial}{\partial r}$, one gets

$$\frac{\partial}{\partial r_i} D_{i\,j}(\mathbf{r}, t) = \frac{\partial}{\partial r} (D_{rr}(r, t) - D_{tt}(r, t)) \frac{r_j}{r} \tag{3.6.17}$$

$$+ \frac{2}{r} (D_{rr}(r, t) - D_{tt}(r, t)) \frac{r_j}{r} + \frac{\partial}{\partial r} D_{tt}(r, t) \frac{r_j}{r} = 0 ,$$

which yields

$$C_{tt}(r, t) = \frac{1}{2r} \frac{\partial}{\partial r} \left(r^2 C_{rr}(r, t) \right) . \tag{3.6.18}$$

The correlation function $C_{i\,j}(\mathbf{r}, t)$ can therefore be described solely in terms of the longitudinal correlation function $C_{rr}(r, t)$.

In summing $C_{i\,j}(\mathbf{r}, t)$ over equal indices i, j, we get $Q_{kin}(r, t)$ that was introduced in the section of the von Kármán-Howarth Eq. (3.3.26) in Eq. (3.3.20)

$$Q_{kin}(r,t) = \sum_{i=j} C_{ij}(\mathbf{r},t) = C_{rr}(r,t) + 2C_{tt}(r,t) = \frac{1}{r^2}\frac{\partial}{\partial r}\left(r^3 C_{rr}(r,t)\right) .$$

$$(3.6.19)$$

The Correlation Function of Third Order

The incompressibility condition for the third-order correlation function,

$$\frac{\partial}{\partial r_k}C_{ij,k}(\mathbf{r},t) = 0 ,$$

$$(3.6.20)$$

has the following implications. Applying (3.6.20) to (3.6.15) yields

$$\left(\frac{1}{r^2}\frac{\partial}{\partial r}\left(r^2 A_1(r,t)\right) + 2r\frac{\partial}{\partial r}\frac{A_3(r,t)}{r}\right)\frac{r_i r_j}{r^2}$$

$$+ \left(\frac{1}{r^2}\frac{\partial}{\partial r}\left(r^2 A_2(r,t)\right) + 2\frac{A_3(r,t)}{r}\right)\delta_{ij} = 0 .$$

$$(3.6.21)$$

Since both brackets in (3.6.21) have to vanish identically in order to satisfy the equation, we get two equations along with (3.6.15) for the three pre-factors $A_1(r,t)$, $A_2(r,t)$ and $A_3(r,t)$. This system of equations is solved by

$$A_1(r,t) = -\frac{r^2}{2}\frac{\partial}{\partial r}\left(\frac{C_{rrr}(r,t)}{r}\right) ,$$

$$(3.6.22)$$

$$A_2(r,t) = -\frac{C_{rrr}(r,t)}{2},$$

$$(3.6.23)$$

$$A_3(r,t) = \frac{1}{4r}\frac{\partial}{\partial r}\left(r^2 C_{rrr}(r,t)\right) .$$

$$(3.6.24)$$

Therefore, the third-order correlation function can be written in terms of $C_{rrr}(r,t)$ only

$$C_{ij,k}(\mathbf{r},t) = -\frac{r^2}{2}\frac{\partial}{\partial r}\left(\frac{C_{rrr}(r,t)}{r}\right)\frac{r_i r_j r_k}{r^3}$$

$$+ \frac{1}{4r}\frac{\partial}{\partial r}\left(r^2 C_{rrr}(r,t)\right)\left(\frac{r_i}{r}\delta_{jk} + \frac{r_j}{r}\delta_{ik}\right) - \frac{C_{rrr}(r,t)}{2}\frac{r_k}{r}\delta_{ij} .$$

$$(3.6.25)$$

References

1. Monin, A.S., Yaglom, A.M.: Statistical Fluid Mechanics: Mechanics of Turbulence. Courier Dover Publications (2007)
2. Onsager, L.: Statistical hydrodynamics. Nuovo Cim. **6**(2), 279–287 (1949)
3. Kolmogorov, A.N.: The local structure of turbulence in incompressible viscous fluid for very large Reynolds numbers. Dokl. Akad. Nauk Sssr **30**(1890), 301–305 (1941)
4. v. Weizsäcker, C.F.: Das Spektrum der Turbulenz bei großen Reynoldsschen Zahlen. Zeitschrift für Phys. **124**(7), 614–627 (1948)
5. Heisenberg, W.: Zur statistischen Theorie der Turbulenz. Zeitschrift für Phys. **124**(7), 628–657 (1948)
6. Faust, G., Argyris, J., Haase, M., Friedrich, R.: An Exploration of Dynamical Systems and Chaos. Springer (2015)
7. Robertson, H.P.: The invariant theory of isotropic turbulence. Math. Proc. Cambridge Philos. Soc. **36**, 209–223 (1940) (Cambridge Univ Press)
8. Chandrasekhar, S.: The theory of axisymmetric turbulence. Philos. Trans. R. Soc. London A Math. Phys. Eng. Sci. **242**(855), 557–577 (1950)
9. Richardson, L.F.: Weather Prediction by Numerical Process. Cambridge University Press (1922)
10. Tennekes, H., Lumley, J.L.: A First Course in Turbulence. MIT Press (1972)
11. Pope, S.B.: Turbulent Flows. Cambridge University Press (2000)
12. Sreenivasan, K.R.: On the universality of the Kolmogorov constant. Phys. Fluids **7**(11), 2778–2784 (1995)
13. Benzi, R., Ciliberto, S., Baudet, C., Chavarria, G.R.: On the scaling of three-dimensional homogeneous and isotropic turbulence. Phys. D Nonlinear Phenom. **80**(4), 385–398 (1995)
14. Yakhot, V.: Mean-field approximation and a small parameter in turbulence theory. Phys. Rev. E **63**, 26307 (2001)
15. Renner, C.: Markowanalysen stochastisch fluktuierender Zeitserien. PhD thesis, Carl von Ossietzky Universität Oldenburg (2002)
16. Renner, C., Peinke, J., Friedrich, R.: Experimental indications for Markov properties of small-scale turbulence. J. Fluid Mech. **433**, 383–409 (2001)
17. Landau, L.D., Lifshitz, E.M.: *Statistical Physics, Third Edition: Volume 5 (Course of Theoretical Physics)*. Butterworth-Heinemann (1987)
18. Kolmogorov, A.N.: A refinement of previous hypotheses concerning the local structure of turbulence in a viscous incompressible fluid at high Reynolds number. J. Fluid Mech. **13**(01), 82–85 (1962)
19. Oboukhov, A.M.: Some specific features of atmospheric tubulence. J. Fluid Mech. **67**(8), 77–81 (1962)
20. Frisch, U.: Turbulence. Cambridge University Press (1995)
21. She, Z.-S., Leveque, E.: Universal scaling laws in fully developed turbulence. Phys. Rev. Lett. **72**(3), 336–339 (1994)
22. Reynolds, O.: An experimental investigation of the circumstances which determine whether the motion of water shall be direct or sinuous, and of the law of resistance in parallel channels. Philos. Trans. R. Soc. London **174**, 935–982 (1883)
23. Taylor, G.I.: Statistical theory of turbulence. *Proc. R. Soc. London A Math. Phys. Eng. Sci.*, 151(873):421–444 (1935)
24. Tropp, E.A., Frenkel, V.Y., Chernin, A.D.: Alexander A. The Man Who Made the Universe Expand. Cambridge University Press, Friedmann (2006)
25. Keller, L.V., Friedman, A.A.: Differentialgleichung für die turbulent Bewegung einer kompressiblen Flüssigkeit. Proc. 1st Intern. Congr. Appl. Delft 395–405 (1924)
26. Landau, L.D., Lifshitz, E.M.: Fluid Mechanics. Butterworth-Heinemann (1987)
27. Prandtl, L.: Führer durch die Strömungslehre. Vieweg (1990)
28. de Karman, T., Howarth, L. On the Statistical Theory of Isotropic Turbulence. Proc. R. Soc. London A Math. Phys. Eng. Sci. **164**(917), 192–215 (1938)

29. McComb, W.D.: The Physics of Fluid Turbulence. Oxford University Press (1990)
30. Lundgren, T.S.: Distribution functions in the statistical theory of turbulence. Phys. Fluids **10**(5), 969–975 (1967)
31. Monin, A.S.: Equations of turbulent motion. J. Appl. Math. Mech. **31**(6), 1057–1068 (1967)
32. Novikov, E.A.: Kinetic Equations for a Vortex Field. Sov. Phys. Dokl. **12**, 1006 (1968)
33. Friedrich, R., Daitche, A., Kamps, O., Lülff, J., Voßkuhle, M., Wilczek, M.: The Lundgren-Monin-Novikov hierarchy: Kinetic equations for turbulence. Compt. R. Phys. **13**(9), 929–953 (2012)
34. Ulinich, F.R., Lyubimov, B.Y.: The statistical theory of turbulence of an incompressible fluid at large Reynolds numbers. Sov. J. Exp. Theor. Phys. **28**, 494 (1969)
35. Hopf, E.: Statistical hydrodynamics and functional calculus. J. Ration. Mech. Anal. **1**(1), 87–123 (1952)
36. Wilczek, M., Daitche, A., Friedrich, R.: Theory for the single-point velocity statistics of fully developed turbulence. Europhys. Lett. **93**(3), 34003 (2011)
37. Wilczek, M., Daitche, A., Friedrich, R.: On the velocity distribution in homogeneous isotropic turbulence: correlations and deviations from Gaussianity. J. Fluid Mech. **676**, 191–217 (2011a)

Chapter 4
Overview of Closure Methods for the Closure Problem of Turbulence

In the preceding chapter, we introduced a statistical formulation of hydrodynamic turbulence. An inherent difficulty in both the moment formulation in Sect. 3.3 and the formulation via kinetic equations in Sect. 3.4 is the hierarchical character of the corresponding system of equations. The latter can be considered as a signature of the spatio-temporal complexity that is inherent in turbulent systems. Despite the fact that moment and kinetic approach differ in the way the hierarchy of equations arises, they share the property that equations of statistical quantities of order n involve unclosed terms of order $n + 1$. Hence, in both approaches, we are faced with the amply defined *closure problem of turbulence.*

The purpose of the present chapter is to give an overview of prevalent concepts that were conceived in order to obtain a self-consistent statistical formulation of turbulence. To this end, we will solely focus on closure assumptions that directly apply to the corresponding hierarchies. Furthermore, it has to be stressed that the following overview of closure assumptions is far from complete and only reflects the author's assessment of their corresponding didactic value.

We will analyze strengths and shortcomings of each closure method and discuss them in chronological order: starting with approximations for the spectral energy transfer, we summarize the famous work of Heisenberg [1] and von Weiszäcker [2] who derived closed equations for the energy spectrum. A similar method will be discussed in the realm of the so-called quasi-normal approximation which neglects certain cumulants of the velocity field. A more sophisticated version of the quasi-normal approximation was given by Orszag [3] in form of the eddy-damped quasi-normal approximation, which was devised to overcome shortcomings in the aforementioned quasi-normal approximation. Subsequently, we will discuss perturbative treatments of the nonlinearity in the Navier-Stokes equation that trace back to Kraichnan [4] and to the more formal diagrammatic method introduced by Wyld [5]. However, we will see that these perturbative methods possess rather informal character, since perturba-

© Springer Nature Switzerland AG 2021
J. Friedrich, *Non-perturbative Methods in Statistical Descriptions of Turbulence*,
Progress in Turbulence - Fundamentals and Applications 1,
https://doi.org/10.1007/978-3-030-51977-3_4

tion expansions are effectuated in powers of the Reynolds number. The latter cannot be considered as a small parameter in fully developed turbulence. Furthermore, we will focus on renormalization group methods from statistical physics and their use in turbulence theory.

4.1 Approximations for the Spectral Energy Transfer

Although Kolmogorov's theory of locally isotropic turbulence concludes with a well-verified functional form of the energy spectrum (3.2.23), it is not derived from "first principles" and thus not sufficient as a closure method (as a matter of fact it contains the unknown pre-factor C_K). However, there are a variety of closure methods for the unknown third-order correlation that appear in the evolution equation of the energy spectrum (3.3.42). An inherent feature of these closure approximations is that they make certain assumptions on how energy is transferred to small scales, i.e., large values of k in Eq. (3.3.42). It is convenient to start from the integrated form of Eq. (3.3.42)

$$\underbrace{\frac{\partial}{\partial t} \int_0^k dk' E(k', t)}_{\substack{\text{energy contained in} \\ \text{large-scale eddies}}}$$

$$= \underbrace{- \int_0^k dk' T(k', t)}_{\substack{\text{spectral energy} \\ \text{transfer } S(k,t)}} \underbrace{- 2\nu \int_0^k dk' k'^2 E(k', t)}_{\substack{\text{energy dissipation rate} \\ \text{within large-scale eddies}}} + \underbrace{\int_0^k dk' Q(k', t)}_{\substack{\text{energy injected into} \\ \text{large-scale eddies}}} . \qquad (4.1.1)$$

The spectral energy transfer $S(k, t)$ through the wavenumber k can be considered as the amount of energy that is transported from macro- to micro-scale per unit time. All subsequently discussed closure approximations of the integrated energy spectrum equation (4.1.1) try to express the spectral energy transfer $S(k, t)$ as a function of the energy spectrum $E(k, t)$.

4.1.1 Kovasznay's Hypothesis

A quite obvious, but rather non-physical closure method was given by Kovasznay [6] on the basis of dimensional analysis of the unclosed spectral energy transfer: If $S(k, t)$ has to be expressed in terms of the energy spectrum $E(k, t)$, it should at least possess the dimensions of an energy flux, i.e.,

$$[S(k,t)] = \frac{[\text{m}]^2}{[\text{s}]^3} \,, \qquad [E(k,t)] = \frac{[\text{m}]^3}{[\text{s}]^2} \,, \qquad [k] = \frac{1}{[\text{m}]} \,. \qquad (4.1.2)$$

Consequently, the transfer function can be expressed according to

$$S(k,t) = 2\alpha_K E(k,t)^{3/2} k^{5/2} \,. \qquad (4.1.3)$$

In this case, it can be shown [7] that the stationary energy spectrum has the form

$$E(k) = \begin{cases} \left(\frac{\langle \varepsilon \rangle}{2\alpha_K}\right)^{2/3} k^{-5/3} \left[1 - \left(\frac{k}{\tilde{k}}\right)\right]^2 & \text{for } k < \tilde{k} \,, \\[2ex] 0 & \text{for } k \geq \tilde{k} \,, \end{cases} \qquad (4.1.4)$$

where we replaced the source term by

$$\int_0^k dk' \, Q(k',t) = \langle \varepsilon \rangle \,, \qquad (4.1.5)$$

and where the cutoff in Eq. (4.1.4) is at $\tilde{k} = 2^{5/4} \alpha_K^{1/2} k_\eta$. Although this closure assumption results in an acceptable shape of the energy spectrum, it only represents a rough approximation of $S(k,t)$ since it totally ignores the influence of eddies belonging to scales $k > \tilde{k}$ and $k < \tilde{k}$. Hence, Kovasznay's closure is non-physical in the sense that it contains no physical mechanism of how energy transfer between different vortical structures effectively takes place. More sophisticated closure assumptions try to find analogies of the spectral energy transfer $S(k,t)$ to certain well-known transport processes such as diffusion processes. Nevertheless, the fact that even the cruel approximation in Eq. (4.1.3) yields reasonable results for the shape of the energy spectrum proves its robustness toward closure approximations.

4.1.2 Heisenberg's Hypothesis

A more instructive closure hypothesis for the integrated spectral equation (4.1.1) was proposed by W. Heisenberg [1] and in a slightly different form in the companion paper by C. F. von Weizsäcker [2]. The closure is based on Boussinesq's approximation of the Reynolds stress in terms of a product of a virtual eddy viscosity and the stress tensor $\frac{\partial u_i}{\partial x_j} + \frac{\partial u_j}{\partial x_i}$ (see also Sect. 3.3.1). The ansatz for the spectral energy transfer reads

$$S(k,t) = 2\nu_e(k,t) \int_0^k dk' k'^2 E(k',t) \,. \qquad (4.1.6)$$

The physical interpretation of this closure assumption is that energy flux toward small scales, represented by $S(k,t)$, acts as an additional dissipative term; it thus models the energy transfer from large-scale vortical structures to fluctuating small-scale vortical structures in the form of a purely diffusive process. Latter has the same character as the conversion of kinetic into thermal energy of the molecular motion in a fluid. Hence, Heisenberg's closure approximation belongs to the family of so-called eddy-viscosity models [8, 9]. Moreover, the influence of the small-scale vortical structures is mimicked by the particular spectral form of the so-called eddy viscosity $v_e(k, t)$ in Eq. (4.1.6), which Heisenberg proposed as

$$v_e(k, t) = \alpha_H \int_k^\infty dk' \sqrt{\frac{E(k', t)}{k'^3}} . \tag{4.1.7}$$

It can be shown [9] that the stationary solution for the energy spectrum $E(k)$ in Eq. (4.1.1) reads

$$E(k) = \left(\frac{8\langle\varepsilon\rangle}{9\alpha_H}\right)^{2/3} k^{-5/3} \left[1 + \frac{8v^3}{3\epsilon\alpha_H^2}k^4\right]^{-4/3} , \tag{4.1.8}$$

which agrees with the Kolmogorov spectrum for intermediate values of k, namely,

$$E(k) = \left(\frac{8\langle\varepsilon\rangle}{9\alpha_H}\right) k^{-5/3} . \tag{4.1.9}$$

For large values of k, i.e., in the dissipation range, Heisenberg's theory predicts the energy spectrum to decay as $E(k) \sim k^{-7}$. This finding stands in contradiction to experimental observations according to experimental findings that suggest an exponential decay for large values of k [7].

4.1.3 Oboukhov's Hypothesis

The earliest closure hypothesis for the equation of the energy spectrum traces back to Oboukhov [10] and was published in the same year as Kolmogorov's theory of locally isotropic turbulence [11] in 1941. The hypothesis is based on the Reynolds decomposition of the velocity field into mean and fluctuating parts, which was discussed in Sect. 3.3.1. Oboukhov assumed that the energy flux is proportional to the Reynolds stress tensor of the fluctuating part multiplied by the velocity gradient of the mean part. Using the same notation as in Sect. 3.3.1, the Reynolds stress tensor of the fluctuating part in Fourier space reads [7]

$$\frac{1}{2} \langle \hat{u}_i'(\mathbf{k})\hat{u}_i'(\mathbf{k})\rangle = \int_k^\infty dk' E(k') . \tag{4.1.10}$$

Moreover, the root mean square of the velocity gradient $\frac{\partial \bar{u}_i(\mathbf{x})}{\partial x_j}$ in Fourier space is proportional to

$$\left[2 \int_0^k dk'^2 E(k') \right]^{1/2} . \tag{4.1.11}$$

Hence, Oboukhov's ansatz for the spectral energy transfer reads

$$S(k) = 2\alpha_O \left[\int_0^k dk'^2 E(k') \right]^{1/2} \int_k^\infty dk'' E(k'') . \tag{4.1.12}$$

Similar to Heisenberg's approximation of the spectral energy transfer, this expression involves disturbances that stem from larger and smaller wavenumbers than the actual scale k, which seems fairly plausible. Nevertheless, Oboukhov's hypothesis splits velocity correlations which belong to the energy flux in a way that leads to non-physical results at large k [7]. The energy spectrum in Oboukhov's approach reads

$$E(k) = \alpha_O^{-3/2} \langle \varepsilon \rangle^{1/4} \nu^{5/4} \begin{cases} 2^{-1/4} \phi'\left(\frac{k}{\tilde{k}}\right) & \text{for } k < \tilde{k} , \\ 0 & \text{for } k \geq \tilde{k} , \end{cases} \tag{4.1.13}$$

where $\phi(x)$ satisfies the cubic equation $4\phi(x)^3 = x^4(1 + \phi(x)^2)$ and $\tilde{k} = 2^{-1/4}\alpha_O^{1/2} k_\eta$. The energy spectrum is thus positive for $k < \tilde{k}$ and falls discontinuously to zero at $k = \tilde{k}$. For small values of $k \ll \tilde{k}$, we obtain $\phi(x) \approx 2^{-2/3}x^{4/3}$ and the energy spectrum reads

$$E(k) = \frac{2^{2/3}}{3} \alpha_O^{-2/3} k^{-5/3} , \tag{4.1.14}$$

in the inertial range. Finally, let us compare the energy spectra that belong to the Kovasznay, Heisenberg, and Oboukhov hypothesis. In order to achieve an overlap in the inertial range of the corresponding spectra, we choose the ratio $\alpha_K : \alpha_H : \alpha_O = \frac{9}{16} : 1 : \frac{\sqrt{3}}{4}$. The energy spectra are plotted in Fig. 4.1. As expected, inertial range spectra follow Kolmogorov's $k^{-5/3}$ law. In the vicinity of the dissipation range, however, the spectra show quite different behavior: In Heisenberg's model, the spectrum continuously goes to zero according to $\sim k^{-7}$. By contrast, Kovasznay's and Oboukhov's predictions drop discontinuously to zero at k_K and k_O, respectively. In this case, the viscous cutoff in Oboukhov's model is located even in front of the dissipation wavenumber k_η. All spectra are in disagreement with experimental findings that indicate exponential decay of the energy spectrum for large k-values.

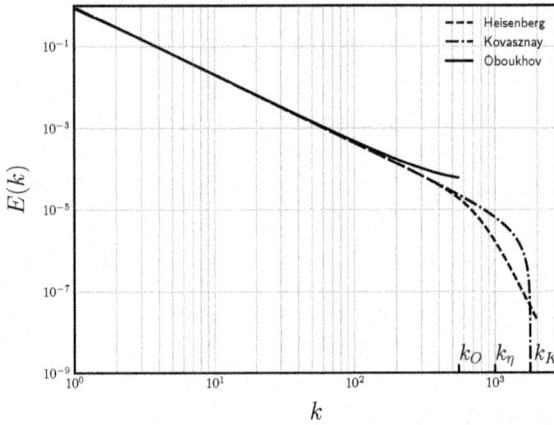

Fig. 4.1 Energy spectra obtained from the stationary energy spectrum equation for the three different closure hypothesis. By the choice $\alpha_K : \alpha_H : \alpha_O = \frac{9}{16} : 1 : \frac{\sqrt{3}}{4}$ in Eqs. (4.1.4), (4.1.6) and (4.1.13) the $k^{-5/3}$-spectra overlap in the inertial range. Further parameters have been chosen according to $\nu = 0.0001$ and $\langle \varepsilon \rangle = 1$. The dissipation range is indicated by k_η and the cutoffs in the Kovasznay and Oboukhov models are indicated by k_K and k_O, respectively

4.2 Quasi-Normal Approximation or Millionschikov's Zero-Fourth Cumulants Hypothesis

Previously discussed closure schemes operate at the stage of the energy spectrum and can thus be considered as two-point closures. These closure hypothesis were introduced on the basis of special physical mechanisms responsible for the energy transfer from large to small scales. A different approach that is of purely statistical nature and bears no direct physical meaning was proposed by Millionschikov in 1941 [12]. He considered a closure at the three-point stage of the Friedmann-Keller hierarchy (3.3.1) that is based on the assumption of vanishing fourth-order cumulant (see Sect. 3.5 for a derivation of the cumulants from the characteristic functional in turbulence). More precisely, Millionschikov suggested that unclosed four-point correlations $\langle u_i(\mathbf{x}, t)u_j(\mathbf{x}', t)u_k(\mathbf{x}'', t)u_l(\mathbf{x}''', t)\rangle$ that appear in the evolution equation of the three-point correlation function are related to moments of second order according to

$$
\begin{aligned}
K_{ijkl} = 0 = {} & \langle u_i(\mathbf{x}, t)u_j(\mathbf{x}', t)u_k(\mathbf{x}'', t)u_l(\mathbf{x}''', t)\rangle \\
& - \langle u_i(\mathbf{x}, t)u_j(\mathbf{x}', t)\rangle \langle u_k(\mathbf{x}'', t)u_l(\mathbf{x}''', t)\rangle \\
& - \langle u_i(\mathbf{x}, t)u_k(\mathbf{x}'', t)\rangle \langle u_j(\mathbf{x}', t)u_l(\mathbf{x}''', t)\rangle \\
& - \langle u_i(\mathbf{x}, t)u_l(\mathbf{x}''', t)\rangle \langle u_k(\mathbf{x}'', t)u_j(\mathbf{x}', t)\rangle \ .
\end{aligned}
\tag{4.2.1}
$$

Although this relation suggests that the velocity field is a Gaussian random field, it should be noted that the *zero-fourth cumulants approximation* makes no assertion on odd order moments which vanish in a purely normal approximation. Hence, purely Gaussian approximations do not admit energy flux across scales, which is a central feature of a turbulent flow.

The fact that the zero-fourth cumulants assumption allows for non-vanishing third-order moments makes it a much weaker assumption in comparison to Gaussian approximations. Therefore, it is often referred to as *quasi-normal approximation*. In order to analyze how the Friedmann-Keller hierarchy can be closed by the quasi-normal hypothesis, it is convenient to work in Fourier space. Here, the unknown term in the equation of the energy spectrum (3.3.42) indeed is the third-order moment (3.3.43). Since the quasi-normal approximation does not specify this moment of odd order, it has to be determined from the subsequent transport equation, namely, Eq. (3.3.40) which couples to correlations of fourth order. The latter can thus be related to moments of second order via the zero-fourth cumulant closure hypothesis in Eq. (4.2.1). For the sake of clarity, we discuss this procedure only at the example of the first nonlinear term in Eq. (3.3.40) since all other terms can be treated in the same manner:

$$
ik_m \left(\delta_{in} - \frac{k_i k_n}{k^2} \right) \int \mathrm{d}\mathbf{k}'' \langle \hat{u}_n(-\mathbf{k} - \mathbf{k}'', t)\hat{u}_m(\mathbf{k}'', t)\hat{u}_j(\mathbf{k}', t)\hat{u}_l(\mathbf{k} - \mathbf{k}', t)\rangle
$$

$$
= ik_m \left(\delta_{in} - \frac{k_i k_n}{k^2} \right) \int \mathrm{d}\mathbf{k}'' \Big[\underbrace{\langle \hat{u}_n(-\mathbf{k} - \mathbf{k}'', t)\hat{u}_m(\mathbf{k}'', t)\rangle}_{\delta(\mathbf{k})\hat{C}_{nm}(-\mathbf{k}-\mathbf{k}'', t)} \langle \hat{u}_j(\mathbf{k}', t)\hat{u}_l(\mathbf{k} - \mathbf{k}', t)\rangle
$$

$$
+ \underbrace{\langle \hat{u}_n(-\mathbf{k} - \mathbf{k}'', t)\hat{u}_j(\mathbf{k}', t)\rangle}_{\delta(-\mathbf{k}-\mathbf{k}''+\mathbf{k}')\hat{C}_{jn}(\mathbf{k}', t)} \langle \hat{u}_m(\mathbf{k}'', t)u_l(\mathbf{k} - \mathbf{k}', t)\rangle
$$

$$
+ \underbrace{\langle \hat{u}_n(-\mathbf{k} - \mathbf{k}'', t)\hat{u}_l(\mathbf{k} - \mathbf{k}', t)\rangle}_{\delta(-\mathbf{k}''-\mathbf{k}')\hat{C}_{ln}(\mathbf{k}+\mathbf{k}'', t)} \langle \hat{u}_j(\mathbf{k}', t)\hat{u}_m(\mathbf{k}'', t)\rangle \Big]
$$

$$
= i \left(\delta_{in}k_m + \delta_{im}k_n - 2\frac{k_i k_m k_n}{k^2} \right) \hat{C}_{jn}(\mathbf{k}', t)\hat{C}_{lm}(\mathbf{k} - \mathbf{k}', t) . \tag{4.2.2}
$$

Here, the term that involves $\delta(\mathbf{k})$ vanishes due to the sifting property of the delta function. Furthermore, we drew on relation (3.3.36). Consequently, Eq. (3.3.40) is approximated according to

$$
\left(\frac{\partial}{\partial t} + \nu k^2 + \nu k'^2 + \nu|\mathbf{k} - \mathbf{k}'|^2 \right) \langle \hat{u}_i(-\mathbf{k}, t)u_j(\mathbf{k}', t)u_l(\mathbf{k} - \mathbf{k}', t)\rangle
$$

$$
= +i \left(\delta_{in}k_m + \delta_{im}k_n - 2\frac{k_i k_m k_n}{k^2} \right) \hat{C}_{jn}(\mathbf{k}', t)\hat{C}_{lm}(\mathbf{k} - \mathbf{k}', t)
$$

$$
-i \left(\delta_{jn}k'_m + \delta_{jm}k'_n - 2\frac{k'_j k'_m k'_n}{k'^2} \right) \hat{C}_{in}(\mathbf{k}, t)\hat{C}_{lm}(\mathbf{k} - \mathbf{k}', t)
$$

$$
-i \left(\delta_{ln}(k_m - k'_m) + \delta_{lm}(k_n - k'_n) - 2\frac{(k_l - k'_l)(k_m - k'_m)(k_n - k'_n)}{|\mathbf{k} - \mathbf{k}'|^2} \right)
$$

$$
\times \hat{C}_{in}(\mathbf{k}, t)\hat{C}_{jm}(\mathbf{k}', t) . \tag{4.2.3}
$$

Integrating this equation yields

$$
\begin{aligned}
\langle \hat{u}_i(-\mathbf{k}, t)\hat{u}_j(\mathbf{k}', t)\hat{u}_l(\mathbf{k} - \mathbf{k}', t)\rangle = &\int_0^t dt' e^{-\nu(k^2 + k'^2 + |\mathbf{k} - \mathbf{k}'|^2)(t - t')} \\
&\times \Bigg[i\left(\delta_{in}k_m + \delta_{im}k_n - 2\frac{k_i k_m k_n}{k^2}\right) \hat{C}_{jn}(\mathbf{k}', t')\hat{C}_{lm}(\mathbf{k} - \mathbf{k}', t') \\
&- i\left(\delta_{jn}k'_m + \delta_{jm}k'_n - 2\frac{k'_j k'_m k'_n}{k'^2}\right) \hat{C}_{in}(\mathbf{k}, t')\hat{C}_{lm}(\mathbf{k} - \mathbf{k}', t') \\
&- i\left(\delta_{ln}(k_m - k'_m) + \delta_{lm}(k_n - k'_n) - 2\frac{(k_l - k'_l)(k_m - k'_m)(k_n - k'_n)}{|\mathbf{k} - \mathbf{k}'|^2}\right) \\
&\times \hat{C}_{in}(\mathbf{k}, t')\hat{C}_{jm}(\mathbf{k}', t')\Bigg].
\end{aligned}
\tag{4.2.4}
$$

This expression has to be inserted for the unknown energy transfer term in the evolution equation of the energy spectrum (3.3.42)

$$
T(k, t) = 2\pi i \sum_{i=j} k_l \int d\mathbf{k}' \langle \hat{u}_i(-\mathbf{k}, t)\hat{u}_j(\mathbf{k}', t)\hat{u}_l(\mathbf{k} - \mathbf{k}', t)\rangle,
\tag{4.2.5}
$$

which yields a closed self-consistent equation for the energy spectrum $E(k, t)$. We want to emphasize that no externally prescribed parameter enters this equation. By contrast, all closures from Sect. 4.1 depend on an additional parameter, which has to be specified in advance.

The disadvantage of the zero-fourth cumulant closure hypothesis is that numerical solutions of the closed Eq. (3.3.42) tend to yield negative values for the energy spectra at intermediate values of k [13]. This is not surprising since it can be shown that setting higher order cumulants to zero necessarily implies that probability density functions cannot be positive everywhere. Nevertheless, there is no reason why an approximate theory should not violate realizability. The problem with quasi-normality is that negative spectra do not constitute a small effect, but are rather pronounced. This implies that the hypothesis is incompatible with the nonlinear dynamics of the Navier-Stokes equation. Furthermore, as it has been pointed out by Orszag [14], splitting fourth-order moments according to Eq. (4.2.1) is responsible for the disproportional buildup of third-order moments. Therefore, he suggested an additional eddy damping term on the l.h.s. of Eq. (4.2.3) and made certain assumptions on time scale separations between eddy damping and inertial terms. This so-called *eddy-damped quasi-normal Markovian approximation* can be considered as a modification of the quasi-normal hypothesis and will be discussed in the following section.

4.3 Eddy-Damped Quasi-Normal Markovian Approximation

In order to avoid the problem of realizability of the quasi-normal approximation discussed in the preceding section, Orszag [14] (see also the monograph by Lesieur [8]) proposed an additional damping term on the l.h.s. of Eq. (4.2.3), namely,

$$
\left(\frac{\partial}{\partial t} + \nu(k^2 + k'^2 + |\mathbf{k} - \mathbf{k}'|^2) + \mu(k, k', |\mathbf{k} - \mathbf{k}'|, t) \right)
$$

$$
\langle \hat{u}_i(-\mathbf{k}, t)\hat{u}_j(\mathbf{k}', t)\hat{u}_l(\mathbf{k} - \mathbf{k}', t) \rangle
$$

$$
= +i \left(\delta_{in}k_m + \delta_{im}k_n - 2\frac{k_i k_m k_n}{k^2} \right) \hat{C}_{jn}(\mathbf{k}', t)\hat{C}_{lm}(\mathbf{k} - \mathbf{k}', t)
$$

$$
-i \left(\delta_{jn}k'_m + \delta_{jm}k'_n - 2\frac{k'_j k'_m k'_n}{k'^2} \right) \hat{C}_{in}(\mathbf{k}, t)\hat{C}_{lm}(\mathbf{k} - \mathbf{k}', t)
$$

$$
-i \left(\delta_{ln}(k_m - k'_m) + \delta_{lm}(k_n - k'_n) - 2\frac{(k_l - k'_l)(k_m - k'_m)(k_n - k'_n)}{|\mathbf{k} - \mathbf{k}'|^2} \right)
$$

$$
\times \hat{C}_{in}(\mathbf{k}, t)\hat{C}_{jm}(\mathbf{k}', t) . \tag{4.3.1}
$$

Latter is motivated by the eddy damping terms involved in approximations of the spectral energy transfer discussed in Sect. 4.1 and has the dimension $[s]^{-1}$. Following the argumentation of Heisenberg's theory 4.1.2, Orszag proposed a damping term

$$
\mu(k, k', |\mathbf{k} - \mathbf{k}'|, t) = \tilde{\mu}(k, t) + \tilde{\mu}(k', t) + \tilde{\mu}(|\mathbf{k} - \mathbf{k}'|, t) , \tag{4.3.2}
$$

where $\tilde{\mu}(k, t) \sim \left[k^3 E(k, t) \right]^{1/2}$. It can be shown that the numerical constant in front of the damping term is associated to the Kolmogorov constant according to $\sim C_k^{3/2}$ [15]. However, this somewhat arbitrary introduction of the damping term in form of Eq. (4.3.2) exhibits contradictory behavior for high values of k if the spectrum is a rapidly decreasing function of k. Under these circumstances, the characteristic frequency of the eddy damping decreases with increasing k. Therefore, a modification of Orszag's damping term that is an increasing function of k has been proposed by Frisch, Lesieur, and Brissaud [16]

$$
\tilde{\mu}(k, t) \sim \left[\int_0^k dk' k'^2 E(k', t) \right]^{1/2} , \tag{4.3.3}
$$

and is influenced by Oboukhov's theory 4.1.3. It thus represents the average rate of deformation of eddies with size $\sim k^{-1}$ by larger eddies. At this point, we must emphasize that the introduction of the eddy damping term at the stage of third-order moments in Eq. (4.3.1) does not lead to a damping of the energy spectrum as it is contained in Heisenberg's and Oboukhov's theory. In fact, the kinetic energy is conserved by nonlinear interactions. The additional damping term on the l.h.s. of Eq.

(4.3.1) results in a modification of the exponential function that enters the third-order moment obtained from the quasi-normal approximation in Eq. (4.2.4)

$$\exp\left[-\nu(k^2 + k'^2 + |\mathbf{k} - \mathbf{k}'|^2)(t - t')\right]$$
$$\rightarrow \exp\left[-[\nu(k^2 + k'^2 + |\mathbf{k} - \mathbf{k}'|^2) + \mu(k, k', |\mathbf{k} - \mathbf{k}'|, t)](t - t')\right] . \quad (4.3.4)$$

Hence, the additional damping term prevents the over-estimation of third-order moments in the equation of the energy spectrum. Nevertheless, in order to guarantee realizability, i.e., non-negativity of the energy spectrum, a further modification is necessary. This so-called *Markovianization* [17] suggests timescale separation between the exponential function and the second-order moments in the integrand of Eq. (4.2.4), namely,

$$\langle \hat{u}_i(-\mathbf{k}, t)\hat{u}_j(\mathbf{k}', t)\hat{u}_l(\mathbf{k} - \mathbf{k}', t)\rangle$$
$$= \int_0^t dt' e^{-[\nu(k^2+k'^2+|\mathbf{k}-\mathbf{k}'|^2)+\mu(k,k',|\mathbf{k}-\mathbf{k}'|,t')](t-t')}$$
$$\times \left[i \left(\delta_{in}k_m + \delta_{im}k_n - 2\frac{k_i k_m k_n}{k^2} \right) \hat{C}_{jn}(\mathbf{k}', t)\hat{C}_{lm}(\mathbf{k} - \mathbf{k}', t) \right.$$
$$- i \left(\delta_{jn}k'_m + \delta_{jm}k'_n - 2\frac{k'_j k'_m k'_n}{k'^2} \right) \hat{C}_{in}(\mathbf{k}, t)\hat{C}_{lm}(\mathbf{k} - \mathbf{k}', t)$$
$$- i \left(\delta_{ln}(k_m - k'_m) + \delta_{lm}(k_n - k'_n) - 2\frac{(k_l - k'_l)(k_m - k'_m)(k_n - k'_n)}{|\mathbf{k} - \mathbf{k}'|^2} \right)$$
$$\left. \times \hat{C}_{in}(\mathbf{k}, t)\hat{C}_{jm}(\mathbf{k}', t) \right] . \quad (4.3.5)$$

The Markovianization assumes that the characteristic time of the inverse argument of the exponential function varies slower than the characteristic time of the products of second-order moments. Furthermore, neglecting the time variations of the eddy damping term yields

$$\int_0^t dt' e^{-[\nu(k^2+k'^2+|\mathbf{k}-\mathbf{k}'|^2)+\mu(k,k',|\mathbf{k}-\mathbf{k}'|)](t-t')}$$
$$= \frac{1 - e^{-[\nu(k^2+k'^2+|\mathbf{k}-\mathbf{k}'|^2)+\mu(k,k',|\mathbf{k}-\mathbf{k}'|)]t}}{\nu(k^2 + k'^2 + |\mathbf{k} - \mathbf{k}'|^2) + \mu(k, k', |\mathbf{k} - \mathbf{k}'|)} . \quad (4.3.6)$$

These assumptions lead to the final form of the *eddy-damped quasi-normal Markovian approximation* (EDQNM). By including the additional eddy damping term, the EDQNM solves the deficiency of negative energy spectra of the quasi-normal approximation. However, the *a posteriori* modifications of the EDQNM slightly shift the focus from pure closure approximations to a more model-like approach to turbulence by the inclusion of an—a priori—unjustified eddy damping term. In the following sections, we will take one step back and discuss a rigorous perturbative treatment of the nonlinearity in the Navier-Stokes equation.

4.4 Renormalization Methods

Before outlining a formal perturbative approach of the nonlinearity in the Navier-Stokes equation (2.1.1), we would like to address similar methods in the field of quantum field theory with the example of quantum electrodynamics. Here, it is possible to treat the interaction term in an adequate way due to the existence of a small parameter, the fine structure constant, which enters the perturbation expansion. On the contrary, turbulence theory exhibits no such small parameter and perturbation expansions are carried out in powers of the Reynolds number, which will be seen in Sect. 4.4.2.

4.4.1 *A Glance at the Concepts from Quantum Field Theory

Quantum field theory is concerned with the generalization of classical field theories, e.g., Maxwell's theory of classical electrodynamics, to a field theory that takes into account the rules of quantum mechanics. In the following, we will mainly focus on a discussion of relevant concepts at the example of quantum electrodynamics (QED), although quantum field theory applies to many other branches in physics (we also refer the reader to the monographs [18], as well as to the introductory texts by Carroll [19] and Huang [20]. The thorough derivation of quantum electrodynamics or quantum field theory, in general, lies beyond the scope of this monograph and we will exclusively outline a rather schematic version of QED. For a more detailed description of QED, we refer the reader to the book by Haken and Wolf [21]. The sole purpose of this section is thus to give a brief overview, in order to retrieve first impressions of why these prevalent concepts of quantum field theory only had limited success in statistical hydrodynamics.

A first step in the derivation of a quantum field theory for classical electrodynamics consists of the quantization of the electromagnetic field [21]. In addition to the usual notion of quantum mechanics that implies quantization of certain observables such as energy and momentum, the so-called *second quantization* leads to the quantization of interacting fields. In case of the interaction between a hydrogen atom and the quantized light field, the process is mediated by the quantized vector potential **A**. Thereby, electromagnetic coupling is determined by the fine structure constant $\alpha \approx \frac{1}{137}$. Due to the smallness of the interaction parameter, QED can be efficiently described as perturbation theory of the electromagnetic quantum vacuum which gives rise to "radiative corrections" of free electron and photon properties. It is important to stress that, within relativistic quantum field theory, photon-electron interaction not only results in fluctuations of the electromagnetic field, but also in fluctuations of the Dirac sea itself. Latter manifest themselves as spontaneous creation and annihilation of virtual electron positron pairs.

In order to avoid confusion between these quantum mechanical effects, we propose to distinguish three basic processes:

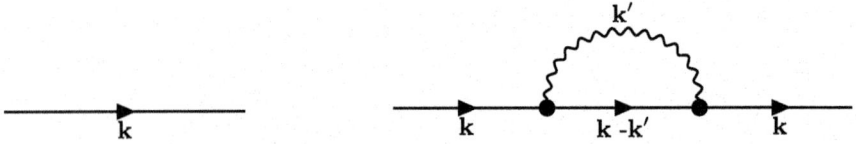

Fig. 4.2 Example of Feynman diagrams: The diagram on the left corresponds to a free electron and can be considered as a transition probability that it goes from a point $\mathbf{x}$ at a time t to a place $\mathbf{x}'$ at a later time t'. It is also referred to as the free electron propagator. The diagram on the right is an example of a virtual emission and subsequent absorption of a photon (wavy line) and represents a first contribution to the self-energy of the electron

- electron self-energy,
- vertex corrections, and
- vacuum polarization.

In the following, we will discuss the first and the last processes separately.

(i.) Electron self-energy

A first instructive consequence of photon-electron interaction is the so-called electron self-energy. It can be best explained with the help of a Feynman diagram. Typical Feynman diagrams are depicted in Fig. 4.2. Here, the left diagram corresponds to the case of the free electron. It can be perceived as the probability amplitude that an electron with momentum $\mathbf{k}$ goes from point $\mathbf{x}$ at time t to point $\mathbf{x}'$ at a later time t'. The entire probability for this process is contained in the square of the probability amplitude which is also referred to as propagator. The Feynman diagram on the right side can be considered as first perturbative influence of the interaction term: first, an electron is situated in its initial condition $\mathbf{k}$ and no photon is present. Then, a photon of momentum $\mathbf{k}'$ is created (wavy line) and the electron transitions into state $\mathbf{k} - \mathbf{k}'$. In the next step, the created photon is absorbed and the electron goes back to its initial state $\mathbf{k}$. Formally, creation and annihilation of the photon arises from a creation and annihilation operator in the matrix element of second-order perturbation theory. Therefore, this and higher order diagrams give rise to the so-called self-energy of the electron. The reason why this procedure is useful is that complicated higher order diagrams give successively smaller contributions to the overall result. More precisely, each vertex (filled circle) introduces a factor α. Therefore, since the coupling constant $\alpha < 1$, the weakly interacting QED theory is amenable to perturbation theory, whereas strongly interacting theories (e.g., quantum chromodynamics) have to be treated in a non-perturbative sense.

So far we omitted a somehow delicate detail in the discussion of the perturbative treatment of the photon-electron interaction: perturbative corrections in the diagram on the right in Fig. 4.2 involve an integration over different field modes $\mathbf{k}'$ of the light field that is divergent due to high $\mathbf{k}'$-modes. This peculiarity leads to the conclusion that one has to deal with an infinite energy shift, which imposed a major problem in the early days of QED. However, latter shortcoming could be overcome by ideas

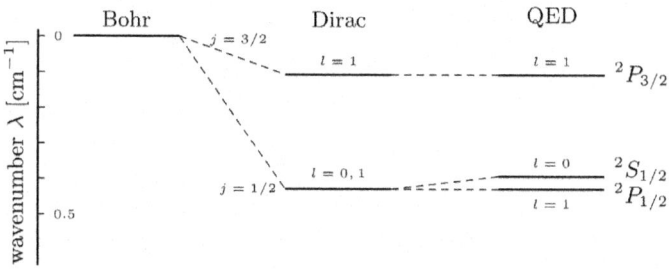

Fig. 4.3 Lamb-Retherford shift of the $n = 2$ —level of the hydrogen atom. Bohr's theory is degenerate with respect to $l = j$. Fine structure splitting occurs in Dirac's theory due to spin-orbit coupling and relativistic corrections which lift the j-degeneracy. However, the $j = 1/2$ —level still is degenerate with respect to $l = 0, 1$. This degeneracy is lifted by photon fluctuations in QED, which lead to an electron mass shift that agrees well with the second-order perturbation theory by Bethe [21]

of Bethe, Schwinger, and Weisskopf who realized that similar calculations for free electrons also lead to infinite results. Therefore, divergent integrals that result from the interaction with the electromagnetic field (Fig. 4.2 right) can be cut off and absorbed into a mass shift of the bare mass of the free electron

> observable mass = bare mass + self-mass . (mass renormalization)

The electromagnetic self-energy can thus be interpreted as a mass shift of the electron from its "bare" mass to its observable mass, which is known as *mass renormalization* [20, 21]. By replacing the bare electron mass with the observable mass, divergent integrals can be cut off at high values of k, which allows for an accurate calculation of the *renormalized energy shift*. Latter was calculated in second-order perturbation theory by Bethe and was measured by Lamb and Retherford [21]. Using high-frequency and microwave spectroscopy, Lamb and Retherford could detect an energy shift between $2^S_{1/2}$ and $2^P_{1/2}$ niveaus of the hydrogen atom depicted in the term diagram in Fig. 4.3. Here, relativistic corrections to Bohr's model due to spin-orbit coupling only lead to lifting of the j-degeneracy, which has been calculated by Dirac [21]. At the time, lifting of the l-degeneracy, which roughly amounts to one part in a million, was not explicable via the relativistic theory of quantum mechanics and could only be understood by continuous absorption and emission of light quanta by the electron of the hydrogen atom in the framework of QED.

(ii.) Vacuum polarization

The notion of the electron self-energy allows for a similar discussion of the photon self-energy. A typical Feynman diagram is depicted in Fig. 4.4. In lowest order, corrections to the free photon propagator (left) manifest themselves by creation and annihilation of a virtual electron-positron pair (right). Since the positron is the anti-

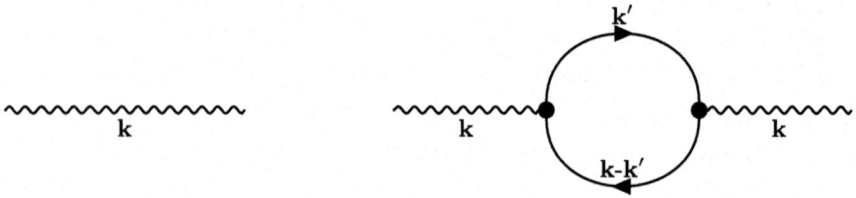

Fig. 4.4 Example of Feynman diagrams for vacuum polarization: The diagram on the left corresponds to a free photon propagator. The diagram on the right shows the lowest order correction to the full propagator. It corresponds to the creation and annihilation of a virtual electron-positron pair. Here, the positron is depicted as an arrow that points backward in time

Fig. 4.5 Charge density of the bare electron (left) and the "dressed" electron (right). The bare electron possesses a delta-peaked charge distribution whereas the charge of the dressed electron is screened by the effect of vacuum polarization

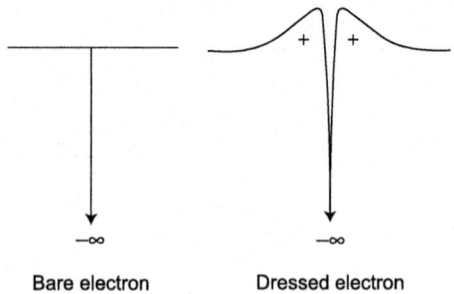

Bare electron   Dressed electron

particle of the electron, its propagator is indicated with an arrow that goes backward in time. The effect of the spontaneous charge separation endows the vacuum with a distribution of induced electric dipoles. Latter effect is referred to as vacuum polarization and results in an alteration of the electron's charge distribution, which is depicted in Fig. 4.5. The alteration of the original charge distribution, a delta peak, leads to an effective charge that is dependent on the distance from the center and can be apprehended in the sense of a Landau quasi-particle, as a "dressed electron": the bare electron charge is screened by vacuum polarizations; however, the screening length is rather small and of the order $\hbar/mc \approx 4 \times 10^{-11}$ cm.

It is interesting to note that in the phenomenon of vacuum polarization, charge renormalization is a universal effect: the screening factor is only dependent on vacuum polarization produced by the photon and is thus the same for all particles. This was shown by Ward [22] and the involved mathematical relations are known as Ward's identities.

Similar to the renormalization procedure in (*i.*) the encountered divergence in the integrations around the loop in Fig. 4.4 can be circumvented via a renormalization of the electron charge

observable charge = bare charge × self charge .   (charge renormalization)

In contrast to mass renormalization, which is additive, charge renormalization is multiplicative.

Concerning higher order corrections to the Feynman diagrams in Figs. 4.2 and 4.4, Dyson [23] showed that all divergent integrals can effectively be absorbed into mass and charge renormalization. Therefore, changing the cutoff solely affects observable parameters such as the charge or mass, whereas the underlying physical mechanisms, represented, for instance, by probability amplitudes, remain the same.

The process of renormalization can be apprehended as some kind of resolving power of the physical system. Therefore, renormalizable theories are necessarily *self-similar*. Hence, at this stage, the use of renormalization methods in turbulence theory is questionable due to the occurrence of non-self-similar statistics. In the following, we will nevertheless describe a formal perturbative treatment of the Navier-Stokes motivated by similar methods to the ones described above.

4.4.2 Primitive Perturbative Treatment of the Nonlinearity in the Navier-Stokes Equation

In this section, we outline a procedure that treats the nonlinear character of the Navier-Stokes equation in a perturbative sense. The treatment implies that the zeroth-order approximation can be obtained from the usual heat equation with an additional stirring force. Perturbation expansions can then be based on the zeroth-order approximation which leads to a switch on of nonlinear interactions, similar to the procedure in QED discussed in the preceding section.

The starting point for the procedure that was developed in the seminal work by Kraichnan [4] is the Navier-Stokes equation in Fourier space

$$\left(\frac{\partial}{\partial t} + \nu k^2 \right) \hat{u}_i(\mathbf{k}, t)$$

$$= M_{ijl}(\mathbf{k}) \int d\mathbf{k}' \hat{u}_j(\mathbf{k}', t) \hat{u}_l(\mathbf{k} - \mathbf{k}', t) + \left(\delta_{ij} - \frac{k_i k_j}{k^2} \right) \hat{f}_j(\mathbf{k}, t) , \quad (4.4.1)$$

where

$$M_{ijl}(\mathbf{k}) = \frac{1}{2i} \left[k_j \left(\delta_{il} - \frac{k_i k_l}{k^2} \right) + k_l \left(\delta_{ij} - \frac{k_i k_j}{k^2} \right) \right] . \quad (4.4.2)$$

The tensorial form in front of the stirring force $\hat{f}_j(\mathbf{k}, t)$ ensures that the velocity field remains solenoidal. For very small Reynolds numbers, the nonlinear term in Eq. (4.4.1) is sub-dominant and hence can be neglected. This implies that inertial interaction forces are small compared to viscous and stirring forces and result in a linear equation for the unperturbed velocity field $\hat{u}_i^{(0)}(\mathbf{k}, t)$ that reads

$$\left(\frac{\partial}{\partial t} + vk^2\right) \hat{u}_i^{(0)}(\mathbf{k}, t) = \left(\delta_{ij} - \frac{k_i k_j}{k^2}\right) \hat{f}_j(\mathbf{k}, t) \,. \tag{4.4.3}$$

This equation can be solved via the method of Green's function according to

$$\hat{u}_i^{(0)}(\mathbf{k}, t) = \int_{-\infty}^{t} \hat{G}_{ij}^{(0)}(\mathbf{k}; t, t') \hat{f}_j(\mathbf{k}, t') \,, \tag{4.4.4}$$

where the tensorial Green's function is given by

$$\hat{G}_{ij}^{(0)}(\mathbf{k}, t, t') = \left(\delta_{ij} - \frac{k_i k_j}{k^2}\right) \hat{G}^{(0)}(k; t, t') \,, \tag{4.4.5}$$

and its defining scalar satisfies

$$\left(\frac{\partial}{\partial t} + vk^2\right) \hat{G}^{(0)}(k; t, t') = \delta(t - t') \,. \tag{4.4.6}$$

Moreover, we can express the unperturbed velocity field at time t in terms of Green's function and the velocity field at an earlier instance in time t' according to

$$\hat{u}_i^{(0)}(\mathbf{k}, t) = \hat{G}^{(0)}(k; t, t')\hat{u}_i^{(0)}(\mathbf{k}, t') \,. \tag{4.4.7}$$

Due to this equation, Green's function $\hat{G}^{(0)}(k; t, t')$ is sometimes referred to as propagator. Furthermore, at this point, it is convenient to transition to the wavenumber-frequency domain where the Navier-Stokes equation takes the form

$$\left(i\omega + vk^2\right) \tilde{u}_i(\mathbf{k}, \omega) \tag{4.4.8}$$

$$= M_{ijl}(\mathbf{k}) \int d\mathbf{k}' \int d\omega' \tilde{u}_j(\mathbf{k}', \omega')\tilde{u}_l(\mathbf{k} - \mathbf{k}', \omega - \omega') + \left(\delta_{ij} - \frac{k_i k_j}{k^2}\right) \tilde{f}_j(\mathbf{k}, \omega) \,.$$

Here tilde signs over the fields $\tilde{u}_i$ indicate their definition in wavenumber-frequency space according to

$$\tilde{u}_i(\mathbf{k}, \omega) = \frac{1}{2\pi} \int dt e^{i\omega t} \hat{u}_i(\mathbf{k}, t) \,. \tag{4.4.9}$$

In addition, Green's function in the wavenumber-frequency space can be obtained from a Fourier transform of Eq. (4.4.6), which yields

$$\tilde{G}^{(0)}(k; \omega) = \frac{1}{i\omega + vk^2} \,. \tag{4.4.10}$$

The following procedure consists of switching on the nonlinear interaction term. In order to apply perturbation theory, higher orders should have diminishing contributions to the power series. Unfortunately, in contrast to the case of QED, the direct use of expansions in terms of the interaction constant will be ineffective since latter is determined by the Reynolds number itself. Therefore, the described perturbative treatment of the nonlinearity in the Navier-Stokes equation has to be considered as a merely informal treatment. Accordingly, the perturbation expansion,

$$\tilde{u}_i(\mathbf{k}, \omega) = \tilde{u}_i^{(0)}(\mathbf{k}, \omega) + \tilde{u}_i^{(1)}(\mathbf{k}, \omega) + \ldots + \tilde{u}_i^{(n)}(\mathbf{k}, \omega) , \qquad (4.4.11)$$

can formally be established. However, its truncation at a given order n is flawed by the fact that higher order contributions are dominant for the case of high Reynolds numbers. In the following, we are interested in the spectral tensor in frequency space

$$\tilde{C}_{ij}(\mathbf{k}; \omega, \omega') = \langle \tilde{u}_i(\mathbf{k}, \omega) \tilde{u}_j(-\mathbf{k}, \omega') \rangle . \qquad (4.4.12)$$

The zeroth-order contribution to this tensor can be obtained from the Fourier transform of Eq. (4.4.4)

$$\tilde{u}_i^{(0)}(\mathbf{k}, \omega) = \left(\delta_{ij} - \frac{k_i k_j}{k^2} \right) \tilde{G}^{(0)}(k; \omega) \tilde{f}_j(\mathbf{k}, \omega) , \qquad (4.4.13)$$

where we made use of the convolution theorem. We thus obtain

$$\tilde{C}_{ij}^{(0)}(\mathbf{k}; \omega, \omega')$$
$$= \langle \tilde{u}_i^{(0)}(\mathbf{k}, \omega) \tilde{u}_j^{(0)}(-\mathbf{k}, \omega') \rangle$$
$$= \left(\delta_{il} - \frac{k_i k_l}{k^2} \right) \left(\delta_{jm} - \frac{k_j k_m}{k^2} \right) \tilde{G}^{(0)}(k; \omega) \tilde{G}^{(0)}(k; \omega') \langle \tilde{f}_l(\mathbf{k}, \omega) f_m(-\mathbf{k}, \omega') \rangle . \qquad (4.4.14)$$

In the following, we want to assume that the stirring force follows a Gaussian distribution with vanishing mean and correlation tensor

$$\langle \tilde{f}_i(\mathbf{k}, \omega) \tilde{f}_j(-\mathbf{k}, \omega') \rangle = (2\pi)^4 \left(\delta_{ij} - \frac{k_i k_j}{k^2} \right) \tilde{\chi}(k; \omega, \omega'). \qquad (4.4.15)$$

In this case, the spectral tensor of order zero reads

$$\tilde{C}_{ij}^{(0)}(\mathbf{k}; \omega, \omega') = (2\pi)^4 \underbrace{\left(\delta_{il} - \frac{k_i k_l}{k^2} \right) \left(\delta_{jm} - \frac{k_j k_m}{k^2} \right) \left(\delta_{lm} - \frac{k_l k_m}{k^2} \right)}_{= \left(\delta_{ij} - \frac{k_i k_j}{k^2} \right)}$$
$$\times \tilde{G}^{(0)}(k; \omega) \tilde{G}^{(0)}(k, \omega') \tilde{\chi}(k; \omega, \omega') . \qquad (4.4.16)$$

The zeroth-order spectral tensor in frequency space can hence be defined in analogy to the spectral tensor (3.3.41)

$$\tilde{C}_{ij}^{(0)}(\mathbf{k}; \omega, \omega') = \left(\delta_{ij} - \frac{k_i k_j}{k^2} \right) E^{(0)}(k; \omega, \omega') , \qquad (4.4.17)$$

with the two-frequency energy spectrum

$$E^{(0)}(k; \omega, \omega') = \tilde{G}^{(0)}(k; \omega)\tilde{G}^{(0)}(k; \omega')\tilde{\chi}(k; \omega, \omega') . \qquad (4.4.18)$$

In general, we are interested in the exact form of $E(k; \omega, \omega')$ in terms of the zeroth-order approximation (4.4.18). Hence, we can insert the perturbation expansion (4.4.11) into the exact spectral tensor in frequency space (4.4.12) and obtain

$$\tilde{C}_{ij}(\mathbf{k}; \omega, \omega') = \left(\delta_{ij} - \frac{k_i k_j}{k^2} \right) E(k; \omega, \omega') \qquad (4.4.19)$$

$$= (2\pi)^4 \left[\left\langle \tilde{u}_i^{(0)}(\mathbf{k}, \omega)\tilde{u}_j^{(0)}(-\mathbf{k}, \omega') \right\rangle + \left\langle \tilde{u}_i^{(2)}(\mathbf{k}, \omega)\tilde{u}_j^{(0)}(-\mathbf{k}, \omega') \right\rangle \right.$$

$$+ \left\langle \tilde{u}_i^{(1)}(\mathbf{k}, \omega)\tilde{u}_j^{(1)}(-\mathbf{k}, \omega') \right\rangle + \left\langle \tilde{u}_i^{(0)}(\mathbf{k}, \omega)\tilde{u}_j^{(2)}(-\mathbf{k}, \omega') \right\rangle + \text{h.o.t.} \Bigg]$$

$$= \left(\delta_{ij} - \frac{k_i k_j}{k^2} \right) E^{(0)}(k; \omega, \omega') + (2\pi)^4 \left[\left\langle \tilde{u}_i^{(2)}(\mathbf{k}, \omega)\tilde{u}_j^{(0)}(-\mathbf{k}, \omega') \right\rangle \right.$$

$$+ \left\langle \tilde{u}_i^{(1)}(\mathbf{k}, \omega)\tilde{u}_j^{(1)}(-\mathbf{k}, \omega') \right\rangle + \left\langle \tilde{u}_i^{(0)}(\mathbf{k}, \omega)\tilde{u}_j^{(2)}(-\mathbf{k}, \omega') \right\rangle + \text{h.o.t.} \Bigg] .$$

In this equation, we already omitted terms like $\tilde{u}_i^{(1)}\tilde{u}_j^{(0)}$ that vanish due to the Gaussianity of the unperturbed velocity field $\tilde{\mathbf{u}}^{(0)}$ which, by Eq. (4.4.13), is a direct consequence of the Gaussianity of the stirring force. Therefore, all higher order terms that are odd functionals of the unperturbed velocity field $\tilde{\mathbf{u}}^{(0)}(-\mathbf{k}, \omega')$ vanish as well. The next step consists of deriving explicit formulas for perturbed coefficients in terms of the unperturbed velocity field directly from the Navier-Stokes equation (4.4.8). To this end, we invert the operator on the l.h.s. of Eq. (4.4.8) and substitute the stirring force by Eq. (4.4.13) which yields

$$\tilde{u}_i(\mathbf{k}, \omega) = \tilde{u}_i^{(0)}(\mathbf{k}, \omega) \qquad (4.4.20)$$

$$+ \tilde{G}^{(0)}(k, \omega)M_{ijl}(\mathbf{k}) \int d\mathbf{k}' \int d\omega' \tilde{u}_j(\mathbf{k}', \omega')\tilde{u}_l(\mathbf{k} - \mathbf{k}', \omega - \omega') .$$

Substituting the perturbation expansion (4.4.11) on both sides of Eq. (4.4.20) yields

$$\tilde{u}_i^{(0)}(\mathbf{k}, \omega) + \tilde{u}_i^{(1)}(\mathbf{k}, \omega) + \tilde{u}_i^{(2)}(\mathbf{k}, \omega) + \ldots = \tilde{u}_i^{(0)}(\mathbf{k}, \omega) \qquad (4.4.21)$$

$$+ \tilde{G}^{(0)}(k, \omega)M_{ijl}(\mathbf{k}) \int d\mathbf{k}' \int d\omega' \left[\tilde{u}_j^{(0)}(\mathbf{k}', \omega')\tilde{u}_l^{(0)}(\mathbf{k} - \mathbf{k}', \omega - \omega') \right.$$

$$+ \tilde{u}_j^{(1)}(\mathbf{k}', \omega')\tilde{u}_l^{(0)}(\mathbf{k} - \mathbf{k}', \omega - \omega') + \tilde{u}_j^{(0)}(\mathbf{k}', \omega')\tilde{u}_l^{(1)}(\mathbf{k} - \mathbf{k}', \omega - \omega') + \ldots \Bigg] .$$

Comparing perturbed coefficients of same order on each site yields

$$\tilde{u}_i^{(0)}(\mathbf{k}, \omega) = \tilde{u}_i^{(0)}(\mathbf{k}, \omega) , \qquad (4.4.22)$$

$$\tilde{u}_i^{(1)}(\mathbf{k}, \omega) = \tilde{G}^{(0)}(k, \omega)M_{ijl}(\mathbf{k}) \int d\mathbf{k}' \int d\omega' \tilde{u}_j^{(0)}(\mathbf{k}', \omega')\tilde{u}_l^{(0)}(\mathbf{k} - \mathbf{k}', \omega - \omega') , \qquad (4.4.23)$$

$$\tilde{u}_i^{(2)}(\mathbf{k}, \omega) = 2\tilde{G}^{(0)}(k, \omega)M_{ijl}(\mathbf{k}) \int d\mathbf{k}' \int d\omega' \tilde{u}_j^{(1)}(\mathbf{k}', \omega')\tilde{u}_l^{(0)}(\mathbf{k} - \mathbf{k}', \omega - \omega') . \qquad (4.4.24)$$

Here, we performed a shift in wavenumber and frequency integrations in the second term of second order in Eq. (4.4.21) which yields the first term of second order, hence the factor of 2 in Eq. (4.4.24). In order to express the second-order coefficient (4.4.24) in terms of the unperturbed velocity field, we insert the perturbed coefficient of first order (4.4.23) and obtain

$$
\tilde{u}_i^{(2)}(\mathbf{k}, \omega) = 2\tilde{G}^{(0)}(k, \omega) M_{ijl}(\mathbf{k}) \int d\mathbf{k}' \int d\omega' \tilde{G}^{(0)}(k', \omega') M_{jmn}(\mathbf{k}')
$$
$$
\times \int d\mathbf{k}'' \int d\omega'' \tilde{u}_m^{(0)}(\mathbf{k}'', \omega'') \tilde{u}_n^{(0)}(\mathbf{k}' - \mathbf{k}'', \omega' - \omega'') \tilde{u}_l^{(0)}(\mathbf{k} - \mathbf{k}', \omega - \omega') .
$$
(4.4.25)

The coefficients (4.4.22–4.4.24) and higher order coefficients allow us to express the spectral tensor in frequency space (4.4.19) in terms of the unperturbed propagator $\tilde{G}^{(0)}(k, \omega)$. The corresponding procedure is explained at the example of the last correlation in Eq. (4.4.19) and is the same for the other two remaining terms of third order

$$
\left\langle \tilde{u}_i^{(0)}(\mathbf{k}, \omega) \tilde{u}_j^{(2)}(-\mathbf{k}, \omega') \right\rangle
$$
$$
= 2\tilde{G}^{(0)}(k, \omega) M_{jlm}(-\mathbf{k}) \int d\mathbf{k}' \int d\omega'' \tilde{G}^{(0)}(k', \omega'') M_{lno}(\mathbf{k}') \int d\mathbf{k}'' \int d\omega'''
$$
$$
\times \underbrace{\left\langle \tilde{u}_i^{(0)}(\mathbf{k}, \omega) \tilde{u}_n^{(0)}(\mathbf{k}'', \omega''') \tilde{u}_o^{(0)}(\mathbf{k}' - \mathbf{k}'', \omega'' - \omega''') \tilde{u}_m^{(0)}(-\mathbf{k} - \mathbf{k}', \omega' - \omega'') \right\rangle}_{\text{Eq.(4.2.1)}}
$$
$$
= 2\tilde{G}^{(0)}(k, \omega) M_{jlm}(-\mathbf{k}) \int d\mathbf{k}' \int d\omega'' \tilde{G}^{(0)}(k', \omega'') M_{lno}(\mathbf{k}') \int d\mathbf{k}'' \int d\omega'''
$$
$$
\left[\left\langle \tilde{u}_i^{(0)}(\mathbf{k}, \omega) \tilde{u}_m^{(0)}(-\mathbf{k} - \mathbf{k}', \omega' - \omega'') \right\rangle \left\langle \tilde{u}_o^{(0)}(\mathbf{k}' - \mathbf{k}'', \omega'' - \omega''') \tilde{u}_n^{(0)}(\mathbf{k}'', \omega''') \right\rangle \right.
$$
$$
\left\langle \tilde{u}_i^{(0)}(\mathbf{k}, \omega) \tilde{u}_n^{(0)}(\mathbf{k}'', \omega''') \right\rangle \left\langle \tilde{u}_o^{(0)}(\mathbf{k}' - \mathbf{k}'', \omega'' - \omega''') \tilde{u}_m^{(0)}(-\mathbf{k} - \mathbf{k}', \omega' - \omega'') \right\rangle
$$
$$
\left. \left\langle \tilde{u}_i^{(0)}(\mathbf{k}, \omega) \tilde{u}_o^{(0)}(\mathbf{k}' - \mathbf{k}'', \omega'' - \omega''') \right\rangle \left\langle \tilde{u}_n^{(0)}(\mathbf{k}'', \omega''') \tilde{u}_m^{(0)}(-\mathbf{k} - \mathbf{k}', \omega' - \omega'') \right\rangle \right]
$$
$$
= 2\tilde{G}^{(0)}(k, \omega) M_{jlm}(-\mathbf{k}) \int d\mathbf{k}' \int d\omega'' \int d\omega''' \tilde{G}^{(0)}(k', \omega'') M_{lno}(\mathbf{k}')
$$
$$
\left[\left(\delta_{in} - \frac{k_i k_n}{k^2} \right) \left(\delta_{om} - \frac{(k_o + k_o')(k_m + k_m')}{|\mathbf{k} + \mathbf{k}'|^2} \right) \right.
$$
$$
\times E^{(0)}(k; \omega, \omega''') E^{(0)}(|\mathbf{k} + \mathbf{k}'|; \omega'' - \omega''', \omega' - \omega'')
$$
$$
\left(\delta_{lo} \frac{k_i k_o}{k^2} \right) \left(\delta_{nm} - \frac{(k_n + k_n')(k_m + k_m')}{|\mathbf{k} + \mathbf{k}'|^2} \right)
$$
$$
\left. \times E^{(0)}(k; \omega, \omega'' - \omega''') E^{(0)}(|\mathbf{k} + \mathbf{k}'|; \omega''', \omega') \right]
$$
$$
= 4\tilde{G}^{(0)}(k, \omega) M_{jlm}(-\mathbf{k}) \int d\mathbf{k}' \int d\omega'' \int d\omega''' G^{(0)}(k', \omega'') M_{lno}(\mathbf{k}')
$$
$$
\left(\delta_{in} - \frac{k_i k_n}{k^2} \right) \left(\delta_{om} - \frac{(k_o + k_o')(k_m + k_m')}{|\mathbf{k} + \mathbf{k}'|^2} \right)
$$
$$
\times E^{(0)}(k; \omega, \omega''') E^{(0)}(|\mathbf{k} + \mathbf{k}'|; \omega'' - \omega''', \omega' - \omega'') .
$$
(4.4.26)

Although the above calculation is rather long, it is worth mentioning two important points: first of all, the quadruple correlation in the third line has been split with the help of the zero-fourth cumulant approximation (4.2.1), an exact relation due to the Gaussianity of the unperturbed velocity field $\tilde{\mathbf{u}}^{(0)}$, which is implied by Eq. (4.4.4). Similar simplifications can be applied to higher order correlations, since cumulants of order higher than two are exactly zero. Hence, the entire procedure reduces itself to the well-known calculus of Gaussian random fields. However, this comes at the cost of questionable truncations of the perturbation expansion in Eq. (4.4.12). Further simplifications in Eq. (4.4.26) arise from the homogeneity property (3.3.36), which also leads to the vanishing of the second term after the splitting by Eq. (4.2.1) due to $M_{lno}(\mathbf{k} = 0) = 0$. Finally, in the last step, we made use of the symmetry $M_{lno} = M_{lon}$ in order to merge the two terms in square brackets into one term. Applying the same procedure to the two remaining terms in Eq. (4.4.12) yields the perturbation expansion

$$
\tilde{C}_{ij}(\mathbf{k}; \omega, \omega') = \left(\delta_{ij} - \frac{k_i k_j}{k^2} \right) E^{(0)}(k; \omega, \omega')
$$

$$
+ 4\tilde{G}^{(0)}(k, \omega) M_{jlm}(-\mathbf{k}) \int d\mathbf{k}' \int d\omega'' \int d\omega''' \tilde{G}^{(0)}(k', \omega'') M_{lno}(\mathbf{k}')
$$

$$
\left(\delta_{in} - \frac{k_i k_n}{k^2} \right) \left(\delta_{om} - \frac{(k_o + k'_o)(k_m + k'_m)}{|\mathbf{k} + \mathbf{k}'|^2} \right)
$$

$$
\times E^{(0)}(k; \omega, \omega''') E^{(0)}(|\mathbf{k} + \mathbf{k}'|; \omega'' - \omega''', \omega' - \omega'')
$$

$$
+ 2\tilde{G}^{(0)}(k, \omega) M_{ilm}(\mathbf{k}) M_{jno}(-\mathbf{k}) \int d\mathbf{k}' \int d\omega'' \int d\omega''' \tilde{G}^{(0)}(k, \omega')
$$

$$
\left(\delta_{lo} - \frac{k'_l k'_o}{k^2} \right) \left(\delta_{mn} - \frac{(k_m - k'_m)(k_n - k'_n)}{|\mathbf{k} - \mathbf{k}'|^2} \right)
$$

$$
\times E^{(0)}(|\mathbf{k} - \mathbf{k}'|; \omega - \omega'', \omega''') E^{(0)}(k'; \omega'', \omega' - \omega''')
$$

$$
+ 4\tilde{G}^{(0)}(k, \omega) M_{ilm}(\mathbf{k}) \int d\mathbf{k}' \int d\omega'' \int d\omega''' \tilde{G}^{(0)}(k', \omega'') M_{lno}(\mathbf{k}')
$$

$$
\left(\delta_{jn} - \frac{k_j k_n}{k^2} \right) \left(\delta_{om} - \frac{(k_o - k'_o)(k_m - k'_m)}{|\mathbf{k} - \mathbf{k}'|^2} \right)
$$

$$
\times E^{(0)}(k; \omega', \omega''') E^{(0)}(|\mathbf{k} - \mathbf{k}'|; \omega'' - \omega''', \omega - \omega'') + \text{h.o.t.} \qquad (4.4.27)
$$

Equation (4.4.27) is formally exact and contains the two-frequency energy spectrum $E^{(0)}(k; \omega, \omega')$, which can be obtained by summing over equal $i = j$. However, the determination of higher order terms in Eq. (4.4.27) is challenging and adds up to a veritable book-keeping task. Moreover, truncating the two-frequency spectrum at a given order is questionable due to the divergence of the series for fully developed turbulence in the high Reynolds number limit. Nevertheless, the following section discusses a diagrammatic representation of the perturbation expansion in a similar fashion as the Feynman diagrams in Sect. 4.4.1. It should be noted that, in contrast to diagrammatic methods encountered in QED, diagrams bear no direct physical meaning. In other words, they do not represent probability amplitudes for generation-recombination processes of particles.

4.4.3 Diagrammatic Representation of the Perturbation Expansion

The diagrams that represent the perturbation coefficients (4.4.22–4.4.24) can be set up via the consideration of the three main constituents, the unperturbed velocity field $\tilde{\mathbf{u}}^{(0)}$, the unperturbed propagator $G^{(0)}$, and the interaction term represented by M_{ijl} [15]. In the following, we will denote them by

- a full line for $\tilde{\mathbf{u}}^{(0)}(\mathbf{k}, \omega)$,
- a broken line for $\tilde{G}^{(0)}(k, \omega)$,
- a (point) vertex for $M_{ijl}(\mathbf{k})$.

The diagram of order zero is obvious and reads

$$\tilde{u}_i^{(0)}(\mathbf{k}, \omega) = \frac{}{\mathbf{k}} \tag{4.4.28}$$

The basic representation of the first-order perturbative coefficient actually consists of the nonlinearity of the Navier-Stokes equation expressed in the unperturbed velocity field

$$\tilde{u}_i^{(1)}(\mathbf{k}, \omega) = \tilde{G}^{(0)}(k, \omega) M_{ijl}(\mathbf{k}) \int d\mathbf{k}' \int d\omega' \tilde{u}_j^{(0)}(\mathbf{k}', \omega') \tilde{u}_l^{(0)}(\mathbf{k} - \mathbf{k}', \omega - \omega')$$

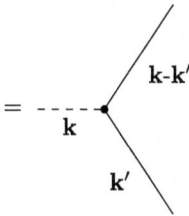

$$\tag{4.4.29}$$

In this representation, the wavenumber at a vertex, i.e., a triadic nonlinear interaction, is conserved from left to right. Wavenumber conservation at vertices applies for higher order coefficients as well, which can be seen from the perturbation coefficient of second order

$$\tilde{u}_i^{(2)}(\mathbf{k}, \omega) = 2\tilde{G}^{(0)}(k, \omega) M_{ijl}(\mathbf{k}) \int d\mathbf{k}' \int d\omega' \tilde{G}^{(0)}(k', \omega') M_{jmn}(\mathbf{k}')$$

$$\times \int d\mathbf{k}'' \int d\omega'' \tilde{u}_m^{(0)}(\mathbf{k}'', \omega'') \tilde{u}_n^{(0)}(\mathbf{k}' - \mathbf{k}'', \omega' - \omega'') \tilde{u}_l^{(0)}(\mathbf{k} - \mathbf{k}', \omega - \omega').$$

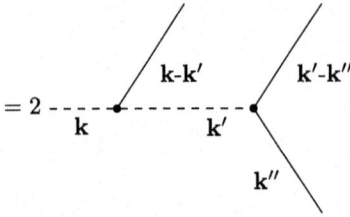

$$(4.4.30)$$

In general, a perturbation diagram of order n is characterized by the occurrence of n vertices corresponding to n triadic interactions. However, in contrast to orders $n \leq 2$, higher order coefficients are composed of different diagrams since vertices can be combined in different ways. The perturbation coefficient of third order, for instance, is composed of two different terms

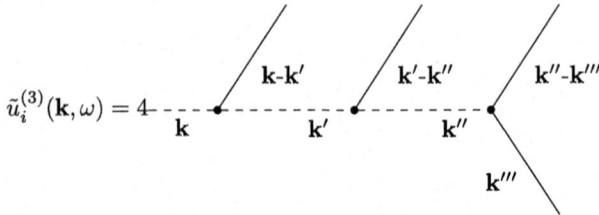

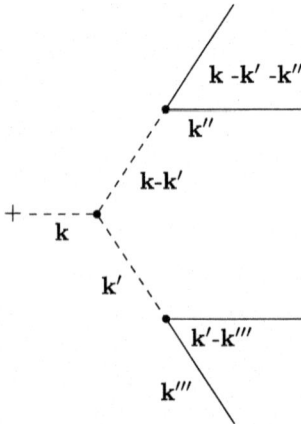

$$(4.4.31)$$

The first term can thus be interpreted as the serial continuation of the second-order diagram, whereas the second term consists of two first-order diagrams that are connected parallelly.

    The further task includes the evaluation of other correlations in the same diagrammatic manner. To this end, we take the corresponding diagrams and their mirror symmetric parts and join them at emergent full lines, i.e., the unperturbed velocity fields $\tilde{\mathbf{u}}^{(0)}$, in all possible ways. The latter procedure corresponds to the factorization of fourth-order moments into two non-vanishing products in Eq. (4.4.26). Next, the evaluation of ensemble averages represented by angular brackets is eventually indicated by a cross at the junction points of the diagrams. Therefore, crosses in diagrams imply correlations of the corresponding emergent velocities $\tilde{\mathbf{u}}^{(0)}$. A first example for this averaging procedure is the correlation of zeroth order

$$\left\langle \tilde{u}_i^{(0)}(\mathbf{k},\omega)\tilde{u}_j^{(0)}(-\mathbf{k},\omega')\right\rangle = \left\langle \underset{\mathbf{k}}{\underline{\qquad}}\ \underset{-\mathbf{k}}{\underline{\qquad}}\right\rangle = \underset{\mathbf{k}}{\underline{\qquad}}\!\times\!\underset{-\mathbf{k}}{\underline{\qquad}} \qquad (4.4.32)$$

At second order, the middle second-order term in Eq. (4.4.27) can be expressed according to

$$\left\langle \tilde{u}_i^{(1)}(\mathbf{k},\omega)\tilde{u}_j^{(1)}(-\mathbf{k},\omega')\right\rangle$$

$$(4.4.33)$$

Here, the factor of 2 arises from the two possible ways in which the full lines of the two diagrams can be connected, in order to obtain duplicate diagrams (4.4.33). This can be attributed to the symmetry $M_{ijn} = M_{inj}$ and can be established for other diagrams as well. For example, we can express the last second-order term that enters in Eq. (4.4.27) according to

$$\left\langle \tilde{u}_i^{(2)}(\mathbf{k},\omega)\tilde{u}_j^{(0)}(-\mathbf{k},\omega')\right\rangle = \left\langle 2\ \cdots \right\rangle$$

$$(4.4.34)$$

The remaining second-order term is simply a mirror symmetric version of diagram (4.4.34). Equation (4.4.35) shows the diagrams for the perturbation expansion

(4.4.27) up to fourth order. Here, we included only four of the 29 possible diagrams of order four.

$$(4.4.35)$$

The particular way in which velocities are correlated in the diagrammatic version of the perturbation expansion (4.4.35) highlights the fact that it involves only two-point (wavenumber) quantities. The diagrammatic average processes exerted in Eqs. (4.4.33–4.4.34) are thus a direct consequence of the Gaussianity of the unperturbed velocity field $\mathbf{u}^{(0)}$. In the following, we will discuss a re-summation technique of the diagrams in Eq. (4.4.35) that was devised by Wyld [5]. For this purpose, it is convenient to introduce new diagrammatic elements, namely,

- a thick full line for the exact velocity field $\mathbf{u}(k, \omega)$,
- a thick broken line for the renormalized propagator $\tilde{G}^{(0)}(k, \omega)$, and
- an open circle for a renormalized vertex.

Moreover, it will be seen that the full correlation tensor (4.4.35) can be divided into two kind of diagrams that we will denote as diagrams of type A and diagrams of type B according to

$$\tilde{C}_{ij}(\mathbf{k}; \omega, \omega') = \tilde{C}_{ij}(\mathbf{k}; \omega, \omega')_A + \tilde{C}_{ij}(\mathbf{k}; \omega, \omega')_B. \qquad (4.4.36)$$

The two types of diagrams in the perturbation series can be distinguished by the following classification:

(i.) Class A diagrams:

Class A diagrams can be split up into two "propagator-like" pieces by cutting an $E^{(0)}(k; \omega, \omega')$-line. In Eq. (4.4.35), e.g., this applies to the zeroth-order diagram itself, the second and third diagrams of the second order and the second and fourth diagrams of fourth order. The entire sum of class A diagrams can thus be absorbed into a renormalized propagator $\tilde{G}(k; \omega')$ that acts on the correlation function of the stirring force.

In order to elucidate this definition, we consider the zeroth-order spectrum in terms of the two Green's functions and the correlation function of the stirring force according to

$$E^{(0)}(k;\omega,\omega') = \tilde{G}^{(0)}(k;\omega)\tilde{\chi}(k;\omega,\omega')\tilde{G}^{(0)}(k;\omega')$$

$$= \mathrm{Tr}\left(\!\!\!\begin{array}{c}\longrightarrow\!\!\ast\!\!\longrightarrow\end{array}\!\!\!\right) = \mathrm{Tr}\left(\!\!\!\begin{array}{c}\text{-}\text{-}\text{-}\text{-}\,\chi\,\text{-}\text{-}\text{-}\text{-}\end{array}\!\!\!\right) \tag{4.4.37}$$

It should be noted that we summed over equal $i = j$ in Eq. (4.4.32) which is designated by the trace symbol of the diagrams. Moreover, the introduction of the new diagram that involves the correlation function of the stirring force has to be understood as a tensor of second order, i.e.,

$$\text{-}\text{-}\text{-}\text{-}\,\chi\,\text{-}\text{-}\text{-}\text{-} = \tilde{G}^{(0)}(k;\omega)\tilde{\chi}(k;\omega,\omega')\tilde{G}^{(0)}(k;\omega')\left(\delta_{ij} - \frac{k_i k_j}{k^2}\right) . \tag{4.4.38}$$

The two class A diagrams of second order can be recast according to

$$(4.4.39)$$

$$(4.4.40)$$

Accordingly, the entire sum of class A diagrams in Eq. (4.4.36) can be rewritten as a decomposition of the form (4.4.37), i.e.,

$$\tilde{C}_{ij}(\mathbf{k};\omega,\omega')_A = \text{-}\text{-}\text{-}\text{-}\,\chi\,\text{-}\text{-}\text{-}\text{-} \tag{4.4.41}$$

where thick dashed lines denote the renormalized propagator $\tilde{G}(k;\omega)$

$$(4.4.42)$$

Equation (4.4.41) thus suggests that all class A diagrams can be obtained via replacing the propagator of zeroth-order $\tilde{G}^{(0)}(k;\omega)$ in Eq. (4.4.38) by the renormalized propagator $\tilde{G}(k;\omega)$ from Eq. (4.4.42).

Let us now turn to the remaining diagrams in the exact spectral tensor (4.4.36) that do not classify as class A diagrams.

(ii.) Class B diagrams:

Class B diagrams are diagrams in Eq. (4.4.35) that cannot be split up into two pieces by cutting an $E^{(0)}(k; \omega, \omega')$-line. In Eq. (4.4.35), e.g., this applies to the first diagram of second order and the first and third diagrams of fourth order. Whereas re-summation of class A diagrams resulted in a renormalized propagator, re-summation of class B diagrams results in a renormalization of the "bare" vertex, i.e., the interaction term.

As an example, we consider the third of the diagrams of order four

$$+8$$

$$(4.4.43)$$

Here, a closer examination of the part

$$4$$

$$(4.4.44)$$

reveals that it connects exactly like a point vertex, i.e., it connects three lines to it. Replacing this part by a bare vertex reduces this diagram to the first diagram of second order (4.4.33). Therefore, it is useful to introduce an expansion of the "full vertex"

$$\circ \; = \; \bullet \; + 4$$

$$(4.4.45)$$

Replacing the bare vertex in the first diagram of second order (4.4.33) by this full vertex reproduces a vast amount of higher order diagrams of type B.

$$2 \;\; = \; 2$$

$$+ \, 8$$

$$(4.4.46)$$

The key to renormalization of class B diagrams can thus be summarized by the following receipt:

Renormalization of class B diagrams:
(a) Find all B-type diagrams that cannot be reduced to a lower order via the replacement of parts by a bare vertex or a propagator of order zero.
(b) Call these diagrams *irreducible diagrams.*
(c) Replace all elements in the irreducible diagrams by their renormalized forms, i.e., expansion (4.4.45) or (4.4.42).
(d) Rewrite these *modified irreducible diagrams* in order to generate a "renormalized" perturbation expansion.

These renormalization procedures for A-type and B-type diagrams can now be used to re-sum Eq. (4.4.35) according to

$$(4.4.47)$$

Obviously, this results in a closed integral equation for the exact spectrum $E(k; \omega, \omega')$. The same renormalization technique is now used to renormalize the expansion of the propagator (4.4.42) and the vertex (4.4.45), which yields

$$(4.4.48)$$

and

$$(4.4.49)$$

The reason for the occurrence of the unrenormalized propagator (thin dashed lines) on the left parts of the diagrams in Eq. (4.4.48) is due to a problem of double-counting of bare diagrams in the primitive perturbation expansion [5]. The renormalization procedure that culminated in the derivation of Eqs. (4.4.47–4.4.49) leads to an effective reduction of the bare perturbation series (4.4.35). The resulting self-consistent integral equations can thus be considered as significant progress compared to the unclosed hierarchy of moments, i.e., the Friedmann-Keller hierarchy.

Nevertheless, we must emphasize that in this case, the described renormalization procedure is not a physical theory in its own right. It has to be considered merely as an appealing reformulation of the moment hierarchy which, due to its divergent unrenormalized series expansion, might prove persistent with regard to truncation. In contrast to effective field theories that are weakly coupled (see the case of QED), perturbative treatment of the strongly interacting degrees of freedom in turbulence, therefore, might end in entirely unreasonable results.

Nonetheless, we want to end this section with the derivation of two well-known theories from the renormalized perturbation expansion. The first one is due to Chandrasekhar [24] and can be obtained from the following truncations:

- Truncate Eq. (4.4.47) at second order (retain two vertices).
- Retain only zeroth-order diagrams in Eqs. (4.4.48) and (4.4.49).

In this case, the equation for the spectral tensor (4.4.47) reduces to

$$\tilde{C}_{ij}(\mathbf{k};\omega,\omega') = \underbrace{}_{} = ----\chi---- + 2 \ ---- \bigcirc ----$$

$$= \left(\delta_{ij} - \frac{k_i k_j}{k^2}\right) \tilde{G}^{(0)}(k;\omega)\tilde{\chi}(k;\omega,\omega')\tilde{G}^{(0)}(k;\omega')$$

$$+ 2\tilde{G}^{(0)}(k,\omega) M_{ilm}(\mathbf{k}) M_{jno}(-\mathbf{k}) \int d\mathbf{k}' \int d\omega'' \int d\omega''' \tilde{G}^{(0)}(k,\omega')$$

$$\times \tilde{C}_{mn}(\mathbf{k}';\omega\omega'',\omega''')\tilde{C}_{lo}(\mathbf{k}-\mathbf{k}';\omega'',\omega'-\omega'''). \tag{4.4.50}$$

This equation is basically the same equation as the one we obtained from the quasi-normal approximation in Sect. 4.2. The second approximative theory that can be recovered via the perturbative treatment is the infamous direct interaction approximation (DIA) by Kraichnan [4]. It obeys the following truncations:

- Truncate Eq. (4.4.47) at second order (retain two vertices).
- Truncate Eq. (4.4.48) at second order.
- Truncate Eq. (4.4.49) at first order, i.e, retain only the bare vertex.

The diagrams that correspond to the DIA read

$$(4.4.51)$$

In the following section, we will discuss similar methods in the realm of the so-called renormalization group in turbulence.

4.5 Renormalization Group Methods

The so-called renormalization group (RG) is applied to the Navier-Stokes equation. The RG, which will be illustrated in the following section, undoubtedly is one of the most successful theories in statistical physics.

4.5.1 *A Glance at the Renormalization Group in Statistical Physics

Before we give a more detailed account of the RG in general, we want to summarize typical features of renormalization which have already been described in the context of quantum electrodynamics in Sect. 4.4.1.

Typical features of renormalization:

(i.) An element of randomness (e.g., fluctuations of the electromagnetic field in QED).
(ii.) Many length and/or time scales.
(iii.) A "bare" quantity is replaced by a "renormalized" quantity.
(iv.) The renormalized quantity typically depends on some "relevant" scale.

Renormalization group methods represent one of the major achievements in statistical physics of the last century [25]. In order to elucidate the reasons for the success of the renormalization group (RG) and to discuss its prevalent concepts, we will focus on the RG in the context of critical phenomena. The latter field deals with matter in the vicinity of a phase transition, for instance, the phase transitions that occur in ferromagnets.

Here, the tendency of magnetic spins to align with each other is opposed by thermal effects which destroys the magnetic order above a critical temperature T_c. Long range interactions are imposed by spin-spin interactions and alignment occurs on all length scales ranging from the lattice constant a up to some correlation length ξ which depends on the temperature T, i.e., $\xi = \xi(T)$. At the critical point T_c, fluctuations occur on all wavelengths from a to $\xi(T_c) \to \infty$. Historically, the renormalization group was first developed by Wilson [26] in pursuance of tackling the phase transition that occurs in the so-called Ising model of a ferromagnet. Although RG methods have entered in nearly every branch of physics, we will mainly focus on its original use in the two-dimensional Ising model. The two-dimensional Ising model is a special case of the Heisenberg model, which is used to describe critical points and phase transitions of magnetic systems. Here, we assume that the spins σ_i are located on a square lattice with lattice constant a and possess only discrete values $\sigma_i = \pm 1$. The Hamiltonian of the two-dimensional Ising model reads

$$\hat{H} = -\sum_{\langle i,j \rangle} J_{ij}\sigma_i\sigma_j + H \sum_i \sigma_i . \qquad (*4.5.1)$$

The first sum in (*4.5.1) includes only nearest neighbors of each spin σ_i whereas the second sum respects the influence of an external magnetic field H. Furthermore, we will assume that spin-spin interaction is homogeneous, which entails constant interaction $J_{ij} = J_0$. Figure 4.7 depicts a quadratic lattice of spins. In the following, the external field H is set to zero. Hence, we restrict ourselves to thermal effects of the spin configuration. For this particular case, there is an analytical way to deduce the partition function,

$$Z = \sum_{\text{all states}} e^{-\hat{H}/k_B T} , \qquad (*4.5.2)$$

of the system [27]. Therefore, the free energy of the system and consequently all other thermodynamic quantities can be obtained from

$$F = -k_B T \ln Z . \qquad (*4.5.3)$$

The two-dimensional Ising model possesses the characteristics of a ferromagnet, i.e., it exhibits a phase transition at the Curie temperature T_c, where it looses its magnetic order. The characteristic of this phase transition is that first derivatives of the free energy are continuous at the critical temperature T_c, whereas second derivatives are discontinuous. This is not the case for a first-order transition, like the liquid-gas transition, where first derivatives are discontinuous [28].

Although the two-dimensional Ising model is exactly solvable, this is not true for the general case, i.e., in three dimensions or in the Heisenberg model that allows for continuous spin values. Therefore, considerable efforts have been devoted to an effective treatment of many-body problems of the form of Eq. (*4.5.1). The famous Landau theory, for instance, is a mean field theory in which any spin can experience a mean field due to the collective behavior of the remaining spins. Under these circumstances and in the vicinity of the critical point, a phenomenological

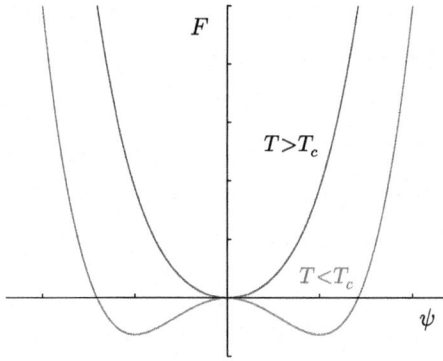

Fig. 4.6 Landau's theory of phase transition of second order in the Ising model. For $T > T_c$, the minimum of the free energy is located at $\psi = 0$, which corresponds to zero total magnetization. The disordered state is thus thermodynamically stable. The phase transition occurs at $T = T_c$, where a new behavior emerges that is characterized by an ordered spin state and therefore a non-zero magnetization

expression for the free energy that respects the symmetry of the Hamiltonian can be derived in the form of a Taylor expansion,

$$F = r\psi^2 + s\psi^4 , \qquad (*4.5.4)$$

of the order parameter ψ that is the coarse-grained field of spins corresponding to the total magnetization. Here, $s > 0$, which guarantees a thermodynamically stable system. As it can be seen in Fig. 4.6, the phase transition occurs if r changes sign, and we can assume that $r = r_0(T - T_c)$. In the case that $T > T_c$, spins are disordered and do not exhibit total magnetization, since the minimum of F is located at $\psi = 0$. For $T < T_c$, however, the disordered phase ϕ becomes unfavorable with respect to the equilibrium magnetization phases $\psi = \pm\sqrt{-r_0(T - T_c)/2s}$. In this approximation, Landau's theory predicts power law behavior in the vicinity of the critical point, for instance, for the magnetization $\sim |T - T_c|^\beta$. However, Landau theory fails to foresee the correct critical exponent β and is therefore an inaccurate description of the collective behavior of spins.

The basic idea of the RG approach is inherited from the general concern of statistical physics, i.e., the reduction of the amount of information necessary to describe a many-body system in going from a micro-state (e.g., 10^{23} individual particles in a gas) to a macro-state description (described by five to six quantities, e.g, pressure, density). The main receipt of the RG is summarized in the following:

(a) (b)

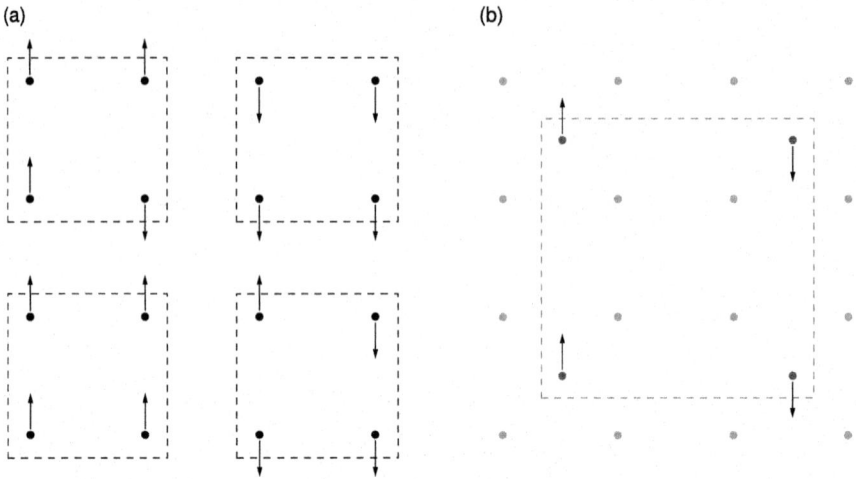

Fig. 4.7 **a** Schematic depiction of the two-dimensional Ising model. The spins are located on a square lattice with lattice constant a. The concept of block spins is indicated by dashed squares. Four spins are repeatedly condensed and can be conceived as a single spin when seen from a larger scale. **b** Coarse-graining of block spins from (**a**) via the majority rule. The new lattice is reduced by a factor of 2 in each direction. Consequently, degrees of freedom are reduced by a factor of 4, from 16 to 4 spins

Simplified procedure of the renormalization group approach:

(i.) Coarse grain our description of the microscopic system.
(ii.) Rescale basic variables such as length scales in order to try to restore the original picture.

In the context of the two-dimensional Ising model, coarse-graining involves the idea of block spins, introduced by Kadanoff [29]. Here, a group of spins, "the block," is assumed to behave like a single spin when viewed from a larger scale. A number of these blocks can then, in turn, be amalgamated to form an even larger block. As indicated in Fig. 4.7, the square lattice is divided into blocks of four spins. The coarse-graining can now be effectuated by the so-called majority rule (in the case of an equal number of like-signed spins the result is left to randomness), although other coarse-graining methods are also possible. The result is depicted in Fig. 4.7b. A new lattice that is shrinked by a factor of two in each direction arises. Therefore, degrees of freedom are reduced by a factor of four. If the system is *scale invariant* then the new lattice is roughly similar to the old one with respect to its physical properties. Mathematically, this corresponds to a fixed point in the sequence of RG transformations.

4.5.1.1 *Renormalization Group Procedure for the 1D Ising Model

In order to give a more mathematical account of the RG procedure, we further restrict ourselves to the one-dimensional Ising model consisting of a long line of spins $\sigma_i = \pm 1$. We can distinguish three extreme situations:

(i.) All dipoles are aligned and the magnetic line has an overall magnetization.
(ii.) The dipoles randomly point up and down and the magnetic line has no overall magnetization.
(iii.) The dipoles alternate between up and down. As in *(i.)*, the system is ordered, however, total magnetization is zero. The latter behavior corresponds to an antiferromagnet and will not be discussed any further.

The probability of any actual state, including the fully aligned state *(i.)* and the unmagnetized state *(ii.)*, is determined by the energy associated with the spin configuration and by the absolute temperature of the system

$$p = \frac{e^{-\hat{H}/k_b T}}{Z} \, . \tag{*4.5.5}$$

Inserting the Hamiltonian from Eq. (*4.5.1) for the case $H = 0$ into the partition function yields

$$Z = \sum_{\text{all states}} \exp\left[K \sum_{\langle i,j\rangle} \sigma_i \sigma_j \right] = \sum_{\text{all states}} \prod_{\langle i,j\rangle} \exp\left[K \sigma_i \sigma_j \right]$$

$$= \sum_{\text{all states}} \prod_{i} \exp\left[K \sigma_i \sigma_{i+1} \right] \, , \tag{*4.5.6}$$

where we introduced the so-called coupling constant $K = J_0/k_b T$ and where the last step was solely a different way of expressing the nearest neighbor property. The above product can also be expressed in terms of even spins only, i.e.,

$$Z = \sum_{\text{all states}} \prod_{i=2,4,\ldots} \exp\left[K \sigma_i (\sigma_{i-1} + \sigma_{i+1}) \right] \, . \tag{*4.5.7}$$

Coarse-graining now consists of partial summation over all even spins $\sigma_{2i} = \pm 1$ which results in a new partition function

$$Z' = \sum_{\ldots,\sigma_1,\sigma_3,\ldots} \prod_{i=2,4,\ldots} \left[e^{K(\sigma_{i-1}+\sigma_{i+1})} + e^{-K(\sigma_{i-1}+\sigma_{i+1})} \right] \, . \tag{*4.5.8}$$

Here, the state summation has to be perceived as a new lattice summation over $\ldots, \sigma_1, \sigma_3, \ldots$, which now includes $N/2$-spins only. By relabeling the spins on the new lattice as $i = 1, 2, \ldots, N/2$, we obtain

$$Z' = \sum_{\text{all states}} \prod_i \left[e^{K(\sigma_{i-1}+\sigma_{i+1})} + e^{-K(\sigma_{i-1}+\sigma_{i+1})} \right] . \qquad (*4.5.9)$$

Now, the invariance condition of the RG transformation demands that the new partition function Z' is similar to the old partition function Z which implies

$$Z' = \sum_{\text{all states}} \prod_i f(K) e^{K'\sigma_i\sigma_j} . \qquad (*4.5.10)$$

Here, the summation includes $N/2$-lattice sites only and a new coupling constant K' has been introduced. Furthermore, we demand that $Z = Z'$ which results in the relation

$$e^{K(\sigma_i+\sigma_{i+1})} + e^{-K(\sigma_i+\sigma_{i+1})} = f(K) e^{K'\sigma_i\sigma_{i+1}} . \qquad (*4.5.11)$$

In the following, we want to derive a relation between the new coupling constant K' and the old one K. To this end, we consider two specific cases:

(i.) $\sigma_i = \sigma_{i+1} = \pm 1$
Eq. (*4.5.11) yields

$$e^{2K} + e^{-2K} = f(K) e^{K'} . \qquad (*4.5.12)$$

(ii.) $\sigma_i = -\sigma_{i+1} = \pm 1$
Eq. (*4.5.11) yields

$$2 = f(K) e^{-K'} . \qquad (*4.5.13)$$

Combining the relations from these two equations yields

$$K' = \frac{1}{2} \ln \cosh(2K) , \qquad (*4.5.14)$$

and a functional form of $f(K)$ that reads

$$f(K) = 2 \cosh^{1/2}(2K) . \qquad (*4.5.15)$$

The relation between old and new partition functions (*4.5.11) can be rewritten as

$$Z(N, K) = f(K)^{N/2} Z(N/2, K') , \qquad (*4.5.16)$$

which is a recursion formula for the partition function. In order to calculate the partition function from Eq. (*4.5.16) recursively, a few further steps are necessary. First of all, since the free energy F is an extensive quantity, we demand that $F(N, K) = Nq(K)$. Furthermore, from Eq. (*4.5.3) we recover

$$\ln Z(N, K) = Nq(K) = N/2 \ln f(K) + \ln Z(N/2, K')$$
$$= N/2 \ln f(K) + N/2q(K') . \qquad (*4.5.17)$$

Rearranging and inserting Eq. (*4.5.15) yields

$$q(K') = 2q(K) - \ln\left[2\cosh^{1/2}(2K)\right] . \qquad (*4.5.18)$$

The final recursion formulas for the partition function can be obtained by inverting this relation and relation (*4.5.14)

$$K = \frac{1}{2}\cosh^{-1}\left(e^{2K'}\right) \quad \text{and} \quad q(K) = \frac{1}{2}\ln 2 + \frac{1}{2}q(K') . \qquad (*4.5.19)$$

In the following, we discuss a numerical evaluation of the partition function. Starting from the high-temperature partition function $Z(N, K) = 2^N$ and the initial guess $K' = 0.01$, we obtain

$$q(K') = \frac{1}{N}\ln Z(N, K') = \frac{1}{N}\ln 2^N = \ln 2 , \qquad (*4.5.20)$$

where we made use of relation $F(N, K) = Nq(K)$. Now, from the first relation in Eq. (*4.5.19) for $K' = 0.01$, we get the old coupling constant $K = 0.100334$. Inserting K into the second relation in Eq. (*4.5.19), we obtain

$$q(K) = \frac{1}{2}\ln 2 + \frac{1}{2} \times 0.01 + \frac{1}{2}\ln 2 . \qquad (*4.5.21)$$

Repeated use of these recursion relations finally leads to the result (to six significant digits)

$$Z(N, K) = [2\cosh K]^N , \qquad (*4.5.22)$$

which is also the exact result. The use of this RG procedure for the one-dimensional Ising model solely leads to the trivial fixed points $K = 0$ or ∞, which is due to the absence of a phase transition. In fact, repeated iteration of the relations (*4.5.19) as it is described above yields the trivial fixed point at infinity. Here, each iteration of the relations (*4.5.19) is a renormalization group transformation which increases the coupling constant and is commonly referred to as *parameter flow*. Figure 4.8 depicts the parameter flow of the coupling constant of the one-dimensional Ising model. However, applying the above method to the two-dimensional Ising model, where a phase transition occurs, yields the non-trivial fixed point at K_c.

$K = 0$ $K = \text{infinity}$

O———▶ ——— ▶— ▶- - - - - ⁓ - - - - - ⁓ - - - — ▶— ▶— ▶ O

High temperature   Low temperature

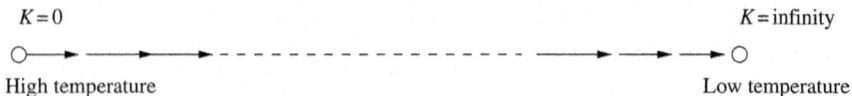

Fig. 4.8 Parameter flow of the coupling constant K in the one-dimensional Ising model. Starting from small coupling constants (random spin configuration) increases the coupling constant and yields the trivial fixed point at infinity

4.5.2 Renormalization Group Methods for the Navier-Stokes Equation

In the following, we want to apply a RG method to the Navier-Stokes equation that has initially been devised by Forster, Stephen, and Nelson [30] and subsequently continued by Yakhot and Orszag [31]. The notations in this section are, however, based on the monograph by McComb [15]. The procedure starts with a Fourier decomposition of the velocity field that is taken to be cutoff for $k > \lambda$

$$u_i(\mathbf{x}, t) = \frac{1}{(2\pi)^4} \int_{k \leq \lambda} d\mathbf{k} \int d\omega e^{i(\mathbf{k}\cdot\mathbf{x} + \omega t)} \tilde{u}_i(\mathbf{k}, \omega) . \qquad (4.5.23)$$

Latter cutoff is also referred to as ultraviolet cutoff. The evolution equation for this decomposition reads

$$\left(i\omega + \nu^{(0)} k^2\right) \tilde{u}_i(\mathbf{k}, \omega)$$
$$= \lambda^{(0)} M_{ijl}(\mathbf{k}) \int_{k' \leq \lambda} d\mathbf{k}' \int d\omega' \tilde{u}_j(\mathbf{k}', \omega)\tilde{u}_l(\mathbf{k} - \mathbf{k}', \omega - \omega')$$
$$+ \left(\delta_{ij} - \frac{k_i k_j}{k^2}\right) \tilde{f}_j(\mathbf{k}, \omega) . \qquad (4.5.24)$$

Here, we already introduced the unrenormalized kinematic viscosity $\nu^{(0)}$ and the parameter $\lambda^{(0)}$ for the unrenormalized interaction term. We will see that the correlation spectrum of stirring forces plays a more central role in the renormalization group approach. Therefore, we will specify their correlation in more detail than before, namely, according to

$$\left\langle \tilde{f}_i(\mathbf{k}, \omega) \tilde{f}_j(\mathbf{k}', \omega')\right\rangle = 2\tilde{\chi}(k)(2\pi)^4 \left(\delta_{ij} - \frac{k_i k_j}{k^2}\right) \delta(\mathbf{k} + \mathbf{k}')\delta(\omega + \omega') , \quad (4.5.25)$$

where the correlation function is specified as a power law in Fourier space

$$\tilde{\chi}(k) = \chi_0 k^{-y} \quad \text{where} \quad \begin{cases} y = -2 & \text{corresponds to thermal equilibrium,} \\ y = 0 & \text{corresponds to macroscopic stirring.} \end{cases}$$

We further divide the velocity into low- and high-wavenumber parts according to

$$
\tilde{u}_i(\mathbf{k}, \omega) = \begin{cases} \tilde{u}_i^<(\mathbf{k}, \omega) & \text{for } 0 < k < \lambda e^{-l} \,, \\ \tilde{u}_i^>(\mathbf{k}, \omega) & \text{for } \lambda e^{-l} < k < \lambda \,, \end{cases} \tag{4.5.26}
$$

which yields the following evolution equations:

$$
\begin{aligned}
\left(i\omega + \nu^{(0)}k^2\right)\tilde{u}_i^<(\mathbf{k}, \omega) &= \left(\delta_{ij} - \frac{k_i k_j}{k^2}\right)\tilde{f}_j^<(\mathbf{k}, \omega) \\
&+ \lambda^{(0)} M_{ijl}^<(\mathbf{k}) \int_{k'\leq\lambda} \mathrm{d}k' \int \mathrm{d}\omega' \left[\tilde{u}_j^<(\mathbf{k}', \omega')\tilde{u}_l^<(\mathbf{k}-\mathbf{k}', \omega-\omega')\right. \\
&+ \left.2\tilde{u}_j^<(\mathbf{k}', \omega')\tilde{u}_l^>(\mathbf{k}-\mathbf{k}', \omega-\omega') + \tilde{u}_j^>(\mathbf{k}', \omega')\tilde{u}_l^>(\mathbf{k}-\mathbf{k}', \omega-\omega')\right],
\end{aligned} \tag{4.5.27}
$$

$$
\begin{aligned}
\left(i\omega + \nu^{(0)}k^2\right)\tilde{u}_i^>(\mathbf{k}, \omega) &= \left(\delta_{ij} - \frac{k_i k_j}{k^2}\right)\tilde{f}_j^>(\mathbf{k}, \omega) \\
&+ \lambda^{(0)} M_{ijl}^>(\mathbf{k}) \int_{k'\leq\lambda} \mathrm{d}k' \int \mathrm{d}\omega' \left[\tilde{u}_j^<(\mathbf{k}', \omega')\tilde{u}_l^<(\mathbf{k}-\mathbf{k}', \omega-\omega')\right. \\
&+ \left.2\tilde{u}_j^<(\mathbf{k}', \omega')\tilde{u}_l^>(\mathbf{k}-\mathbf{k}', \omega-\omega') + \tilde{u}_j^>(\mathbf{k}', \omega')\tilde{u}_l^>(\mathbf{k}-\mathbf{k}', \omega-\omega')\right].
\end{aligned} \tag{4.5.28}
$$

Renormalization group method in turbulence:
(a) Eliminate high-$\mathbf{k}$ modes by solving the evolution equation for $\tilde{\mathbf{u}}^>$ (4.5.28) and substitute the solution into the evolution equation for low-k modes $\tilde{\mathbf{u}}^<$ (4.5.27). Take the average over $\tilde{\mathbf{f}}^>$.
(b) Rescale $\mathbf{k}, t, \tilde{\mathbf{u}}^<(\mathbf{k}, t)$ and $\tilde{\mathbf{f}}^>$ in such a manner that the new equations look like the original Navier-Stokes equation.

At this point, we must emphasize that in the original work of Forster, Nelson, and Stephen [30], the cutoff λ was chosen to lie well inside the dissipation range, i.e., it entirely excluded inertial interactions. Despite the fact that this theory thus not qualifies as adequate statistical theory of turbulence, it is quite informative in the sense that it allows for a renormalization of viscosity, stirring forces, and coupling constants. Moreover, several attempts have been made to generalize this efficient method to higher wavenumber cutoffs [31]. However, it must be stressed that the effective coupling constant increases with an increasing wavenumber cutoff. In order to discuss the basic steps, we will follow the original theory in assuming that $\lambda \ll \eta$. To this end, we establish a perturbative treatment of the high-wavenumber velocity field $\tilde{\mathbf{u}}^>$ along the lines of Sect. 4.4.2. Hence, we are interested in a perturbation expansion of the form

$$
\tilde{u}_i^>(\mathbf{k}, \omega) = \tilde{u}_i^{>(0)}(\mathbf{k}, \omega) + \lambda^{(0)}\tilde{u}_i^{>(1)}(\mathbf{k}, \omega) + \ldots (\lambda^{(0)})^n \tilde{u}_i^{>(n)}(\mathbf{k}, \omega) \,. \tag{4.5.29}
$$

Obviously, an identical series expansion can be established for the low-wavenumber velocity field $\tilde{\mathbf{u}}^<$, but here we are only interested in a perturbative solution for $\tilde{\mathbf{u}}^>$ in order to eliminate its effects in the evolution equation of the large scales (4.5.27). Latter procedure can thus be considered as an effective coarse-graining of solutions. Inserting the perturbation expansion (4.5.29) into Eq. (4.5.28) and comparing orders yield

$$\tilde{u}_i^{>(0)}(\mathbf{k}, \omega) = \tilde{G}^{(0)}(k, \omega)\tilde{f}_i^>(\mathbf{k}, \omega) , \qquad (4.5.30)$$

for zeroth order and

$$\tilde{u}_i^{>(1)}(\mathbf{k}, \omega) \qquad\qquad\qquad\qquad\qquad\qquad\qquad (4.5.31)$$
$$= \tilde{G}^{(0)}(k, \omega)\lambda^{(0)} M_{ijl}^>(\mathbf{k}) \int_{k' \leq \lambda} d\mathbf{k}' \int d\omega' \Big[\tilde{u}_j^<(\mathbf{k}', \omega')\tilde{u}_l^<(\mathbf{k} - \mathbf{k}', \omega - \omega')$$
$$+ 2\tilde{u}_j^<(\mathbf{k}', \omega')\tilde{u}_l^{>(0)}(\mathbf{k} - \mathbf{k}', \omega - \omega') + \tilde{u}_j^{>(0)}(\mathbf{k}', \omega')\tilde{u}_l^{>(0)}(\mathbf{k} - \mathbf{k}', \omega - \omega')\Big] ,$$

$$\tilde{u}_i^{>(2)}(\mathbf{k}, \omega)$$
$$= 2\tilde{G}^{(0)}(k, \omega)\lambda^{(0)} M_{ijl}^>(\mathbf{k}) \int_{k' \leq \lambda} d\mathbf{k}' \int d\omega' \Big[2\tilde{u}_j^<(\mathbf{k}', \omega')\tilde{u}_l^{>(1)}(\mathbf{k} - \mathbf{k}', \omega - \omega')$$
$$+ 2\tilde{u}_j^{>(0)}(\mathbf{k}', \omega')\tilde{u}_l^{>(1)}(\mathbf{k} - \mathbf{k}', \omega - \omega')\Big] , \qquad (4.5.32)$$

for first and second orders. The perturbation coefficient $\tilde{\mathbf{u}}^{>(1)}$ that enters in the last equation (4.5.32) can be expressed in terms of the unperturbed coefficient via (4.5.32) which is not exerted here, but yields a similar expression as Eq. (4.4.25) from Sect. 4.4.2. Inverting the operator on the l.h.s. of the evolution equation for the large-scale velocity field (4.5.27) and inserting the perturbation expansion for the small-scale velocity field (4.5.29) yield

$$\tilde{u}_i^<(\mathbf{k}, \omega) = \left(\delta_{ij} - \frac{k_i k_j}{k^2}\right) \tilde{G}^{(0)}(k, \omega)\tilde{f}_j^<(\mathbf{k}, \omega)$$
$$+ \lambda^{(0)} M_{ijl}^<(\mathbf{k}) \int_{k' \leq \lambda} d\mathbf{k}' \int d\omega' \Big[\tilde{u}_j^<(\mathbf{k}', \omega')\tilde{u}_l^<(\mathbf{k} - \mathbf{k}', \omega - \omega') + 2\tilde{u}_j^<(\mathbf{k}', \omega')$$
$$\Big\{\tilde{u}_l^{>(0)}(\mathbf{k} - \mathbf{k}', \omega - \omega') + \lambda^{(0)}\tilde{u}_l^{>(1)}(\mathbf{k} - \mathbf{k}', \omega - \omega') + (\lambda^{(0)})^2\tilde{u}_l^{>(2)}(\mathbf{k} - \mathbf{k}', \omega - \omega')\Big\}$$
$$+ \tilde{u}_j^{>(0)}(\mathbf{k}', \omega')\tilde{u}_l^{>(0)}(\mathbf{k} - \mathbf{k}', \omega - \omega') + \lambda^{(0)}\tilde{u}_j^{>(0)}(\mathbf{k}', \omega')\tilde{u}_l^{>(1)}(\mathbf{k} - \mathbf{k}', \omega - \omega')$$
$$+ \lambda^{(0)}\tilde{u}_j^{>(1)}(\mathbf{k}', \omega')\tilde{u}_l^{>(0)}(\mathbf{k} - \mathbf{k}', \omega - \omega') + (\lambda^{(0)})^2\tilde{u}_j^{>(1)}(\mathbf{k}', \omega')\tilde{u}_l^{>(1)}(\mathbf{k} - \mathbf{k}', \omega - \omega')$$
$$+ (\lambda^{(0)})^2\tilde{u}_j^{>(0)}(\mathbf{k}', \omega')\tilde{u}_l^{>(2)}(\mathbf{k} - \mathbf{k}', \omega - \omega')$$
$$+ (\lambda^{(0)})^2\tilde{u}_j^{>(2)}(\mathbf{k}', \omega')\tilde{u}_l^{>(0)}(\mathbf{k} - \mathbf{k}', \omega - \omega')\Big] + \mathcal{O}((\lambda^{(0)})^4) . \qquad (4.5.33)$$

We can now calculate the average of high-wavenumber modes (small scales) via the following rules, where $\langle \cdots \rangle_>$ denotes the high-wavenumber average.

Averaging procedure for the high-wavenumber (small-scale) velocity field that enters in Eq. (4.5.33):

a) Statistical independence of the low-wavenumber (large-scale) velocity field $\tilde{\mathbf{u}}^<$ from the high-wavenumber modes, i.e., $\langle \tilde{\mathbf{u}}^< \rangle_> = \tilde{\mathbf{u}}^<$ and $\langle \tilde{\mathbf{f}}^< \rangle_> = \tilde{\mathbf{f}}^<$.

b) Homogeneity of stirring forces that implies $M_{ijl}^>(\mathbf{k})\langle \tilde{u}_j^{>(0)}(\mathbf{k}', \omega)\tilde{u}_l^{>(0)}(\mathbf{k} - \mathbf{k}', \omega - \omega')\rangle_> = 0$, since $M_{ijl}^>(\mathbf{k} = 0) = 0$.

c) Stirring forces have zero mean which yields $\langle \tilde{\mathbf{u}}^{>(0)} \rangle_> = 0$ and $\langle \tilde{\mathbf{f}}^> \rangle_> = 0$.

d) Stirring forces follow a Gaussian distribution.

e) Averages of the high-wavenumber (small-scale) velocity field $\tilde{\mathbf{u}}^>$ can be performed in making use of Eq. (4.5.30), i.e., under the assumption of the Gaussianity of the unperturbed velocity field $\tilde{\mathbf{u}}^{>(0)}$, see point d).

Eq. (4.5.33) can now be simplified via this averaging procedure according to

$$
\tilde{u}_i^<(\mathbf{k}, \omega) = \left(\delta_{ij} - \frac{k_i k_j}{k^2} \right) \tilde{G}^{(0)}(k, \omega) \tilde{f}_j^<(\mathbf{k}, \omega)
$$

$$
+ \tilde{G}^{(0)}(k, \omega) \lambda^{(0)} M_{ijl}^<(\mathbf{k}) \int_{k \leq \lambda} d\mathbf{k}' \int d\omega' \tilde{u}_j^<(\mathbf{k}', \omega) \tilde{u}_l^<(\mathbf{k} - \mathbf{k}', \omega - \omega')
$$

$$
+ 2 \tilde{G}^{(0)}(k, \omega) (\lambda^{(0)})^2 M_{ijl}^<(\mathbf{k}) \int_{k' \leq \lambda} d\mathbf{k}' \int d\omega' \tilde{G}^{(0)}(|\mathbf{k} - \mathbf{k}'|, \omega - \omega')
$$

$$
\times M_{lmn}^>(\mathbf{k} - \mathbf{k}') \int_{k'' \leq \lambda} d\mathbf{k}'' \int d\omega'' \tilde{u}_j^<(\mathbf{k}', \omega') \tilde{u}_m^<(\mathbf{k}'', \omega'') \tilde{u}_n^<(\mathbf{k}' - \mathbf{k}'', \omega' - \omega'')
$$

$$
+ 8(2\pi)^4 (\lambda^{(0)})^2 M_{mjl}^<(\mathbf{k}) \int_{k' \leq \lambda} d\mathbf{k}' \int d\omega' \tilde{G}^{(0)}(|\mathbf{k} - \mathbf{k}'|, \omega - \omega') \tilde{G}^{(0)}(k', \omega')^2
$$

$$
\times M_{lnm}^<(\mathbf{k} - \mathbf{k}') \tilde{\chi}(k') \left(\delta_{jn} - \frac{k_j' k_n'}{k'^2} \right) \tilde{u}_i^<(\mathbf{k}, \omega) + \mathcal{O}((\lambda^{(0)})^3) . \tag{4.5.34}
$$

In rearranging this equation, i.e., by multiplying with $(i\omega + \nu^{(0)}k^2)$, we can see that the last term acts as an additional viscous term and we can recover an equation that is fairly similar to Eq. (4.5.27), namely,

$$
\left(i\omega + \nu^{(0)}k^2 + \Delta\nu^{(0)}(k)k^2 \right) \tilde{u}_i^<(\mathbf{k}, \omega)
$$

$$
= M_{ijl}^<(\mathbf{k}) \int_{k' \leq \lambda} d\mathbf{k}' \int d\omega' \tilde{u}_j^<(\mathbf{k}', \omega) \tilde{u}_l^<(\mathbf{k} - \mathbf{k}', \omega - \omega') M_{ijl}^<(\mathbf{k})
$$

$$
\int_{k' \leq \lambda} d\mathbf{k}' \int d\omega' \tilde{G}^{(0)}(|\mathbf{k} - \mathbf{k}'|, \omega - \omega') M_{lmn}^>(\mathbf{k} - \mathbf{k}') \int_{k'' \leq \lambda} d\mathbf{k}'' \int d\omega''
$$

$$
\tilde{u}_j^<(\mathbf{k}', \omega') \tilde{u}_m^<(\mathbf{k}'', \omega'') \tilde{u}_n^<(\mathbf{k}' - \mathbf{k}'', \omega' - \omega'') + \left(\delta_{ij} - \frac{k_i k_j}{k^2} \right) \tilde{f}_j^<(\mathbf{k}, \omega) .
$$

$$
\tag{4.5.35}
$$

The similarity to Eq. (4.5.27) becomes even more apparent if we consider the limit $k \to 0$ in Eq. (4.5.35) that allows for the neglect of the triple product of $\tilde{\mathbf{u}}^<$ in the last term. In this case, we obtain an iteration equation of the form

$$
\left(i\omega + \nu^{(0)} k^2 + \Delta \nu^{(0)}(k) k^2 \right) \tilde{u}_i^<(\mathbf{k}, \omega) \tag{4.5.36}
$$
$$
= M_{ijl}^<(\mathbf{k}) \int_{k' \leq \lambda} d\mathbf{k}' \int d\omega' \, \tilde{u}_j^<(\mathbf{k}', \omega) \tilde{u}_l^<(\mathbf{k} - \mathbf{k}', \omega - \omega') + \left(\delta_{ij} - \frac{k_i k_j}{k^2} \right) \tilde{f}_j^<(\mathbf{k}, \omega),
$$

where the effective viscosity reads

$$
\Delta \nu^{(0)}(k) = 8(2\pi)^4 k^{-2} M_{mjl}^<(\mathbf{k}) \int_{k' \leq \lambda} d\mathbf{k}' \int d\omega' \tag{4.5.37}
$$
$$
\times \, \tilde{G}^{(0)}(|\mathbf{k} - \mathbf{k}'|, \omega - \omega') \tilde{G}^{(0)}(k', \omega')^2 M_{lnm}^<(\mathbf{k} - \mathbf{k}') \tilde{\chi}(k') \left(\delta_{jn} - \frac{k_j' k_n'}{k'^2} \right).
$$

The previous discussion that treated a three-dimensional turbulent velocity field can be generalized to d dimensions as well. In order to be more general, we will give here the d-dimensional form of the effective viscosity that can be calculated explicitly by integrating over angles in d-dimensional wavenumber space and then letting $k \to 0$, which yields

$$
\Delta \nu^{(0)}(0) = \frac{K(d)(\lambda^{(0)})^2 \chi_0}{(\nu^{(0)})^2 \lambda^\epsilon} \frac{\exp(\epsilon l) - 1}{\epsilon}, \tag{4.5.38}
$$

where

$$
\epsilon = 4 + y - d, \qquad K(d) = \frac{A(d) S_d}{(2\pi)^d}, \quad \text{and} \quad A(d) = \frac{d^2 - d - \epsilon}{2d(d+2)}. \tag{4.5.39}
$$

Furthermore, S_d is the area of the d-dimensional unit sphere

$$
S_d = \frac{2\pi^{d/2}}{\Gamma(d/2)}. \tag{4.5.40}
$$

Therefore, the perturbed viscosity reads

$$
\nu^{(1)} = \nu^{(0)} + \Delta \nu^{(0)}(0) = \nu^{(0)} \left(1 + K(d) \left(\lambda^{(1)} \right)^2 \frac{\exp(\epsilon l) - 1}{\epsilon} \right), \tag{4.5.41}
$$

with the modified coupling constant

$$\lambda^{(1)} = \frac{\lambda^{(0)} \chi_0^{1/2}}{\left(\nu^{(0)}\right)^{3/2} \lambda^{\epsilon/2}} , \tag{4.5.42}$$

and the renormalized propagator

$$\tilde{G}^{(1)}(k, \omega) = \frac{1}{i\omega + \nu^{(1)} k^2} . \tag{4.5.43}$$

In the next steps, it is our objective to rescale the iterative equation (4.5.35) in order to make it even more similar to the original Navier-Stokes equation. First of all, since Eq. (4.5.35) only is defined on the wavenumber interval $0 < k < \lambda e^{-l}$, we have to coarsen it by the introduction of the new variables

$$\bar{k} = k e^{-l} \qquad \bar{\omega} = \omega e^{a(l)} \qquad \bar{u}_i(\bar{\mathbf{k}}, \bar{\omega}) = \tilde{u}_i^<(\mathbf{k}, \omega) e^{-c(l)} . \tag{4.5.44}$$

Here, rescaling of frequency and velocity field has been introduced in order to keep the term $i\omega \tilde{u}_i^<(\mathbf{k}, \omega)$ invariant under the wavenumber rescaling relation. Moreover, the rescaling parameters $a(l)$ and $c(l)$ have yet to be determined. Applying the renormalization group transformations (4.5.44) to Eq. (4.5.35) yields

$$\left(i\bar{\omega} + \bar{\nu}\bar{k}^2\right) \bar{u}_i(\mathbf{k}, \bar{\omega}) + \bar{\lambda} M_{ijl}(\bar{\mathbf{k}}) \int_{\bar{k}' \le \lambda} d\bar{\mathbf{k}}' \int d\bar{\omega}' \bar{u}_j(\bar{\mathbf{k}}', \bar{\omega}) \tilde{u}_l^<(\bar{\mathbf{k}} - \bar{\mathbf{k}}', \bar{\omega} - \bar{\omega}')$$

$$= \left(\delta_{ij} - \frac{\bar{k}_i \bar{k}_j}{\bar{k}^2}\right) \bar{f}_j(\mathbf{k}, \omega) , \tag{4.5.45}$$

where the rescaled force, viscosity, and coupling constant read

$$\bar{f}_i(\bar{\mathbf{k}}, \bar{\omega}) = \tilde{f}_i^<(\mathbf{k}, \omega) e^{a(l)-c(l)}, \tag{4.5.46}$$

$$\bar{\nu} = \nu^{(1)} e^{a(l)-2c(l)}, \tag{4.5.47}$$

$$\bar{\lambda} = \lambda^{(0)} e^{c(l)-(d(l)+1)l} , \tag{4.5.48}$$

where the dimension d has been introduced in order to be more general than the special case $d = 3$ in Eq. (4.5.45). Although the stirring force is also subject to rescaling, its energy transfer to the system has to be unaffected by the rescaling relation, i.e.,

$$\left\langle \tilde{f}_i(\mathbf{k}, \omega) \tilde{f}_j(\mathbf{k}', \omega') \right\rangle = 2 \tilde{\chi}(k)(2\pi)^4 \left(\delta_{ij} - \frac{k_i k_j}{k^2}\right) \delta(\mathbf{k} + \mathbf{k}') \delta(\omega + \omega') \tag{4.5.49}$$

has to be identical to Eq. (4.5.25). Therefore, we obtain an important first relation between the rescaling parameters $a(l)$ and $c(l)$ that reads

$$2c(l) = 3a(l) + (y + d)l . \tag{4.5.50}$$

In the original work [30], the iteration of Eq. (4.5.35) was carried out for infinitesimal wavenumber bands at each step. Therefore, recursion relations for the viscosity (4.5.41) and for the coupling constant (4.5.42) become ordinary differential equations for continuous l

$$\frac{d\nu(l)}{dl} = \nu(l)(z(l) - 2 + K(d)\lambda(l)) , \qquad (4.5.51)$$

$$\frac{d\chi_0(l)}{dl} = 0 , \qquad (4.5.52)$$

$$\frac{d\lambda(l)}{dl} = \frac{\lambda(l)}{2} \left(\epsilon - 3K(d)\lambda(l)^2\right) , \qquad (4.5.53)$$

where $z(l) = \frac{da(l)}{dl}$. Since the last of these differential equations determines the strength of the interaction term, it is quite interesting to investigate the behavior of the coupling constant $\lambda(l)$ in Eq. (4.5.53) for varying ϵ. However, we must keep in mind that the above procedure solely applies for the case $k \to 0$, i.e., its meaning for fully developed turbulence is rather residual. Nevertheless, in observing the particular form of Eq. (4.5.53) it can be seen that the renormalized coupling constant exhibits a supercritical pitchfork bifurcation at $\epsilon = 0$. This implies that for $\epsilon < 0$ we have a stable fixed point $\lambda_1^* = 0$ and $\lambda(l)$ grows exponentially to λ_1^*. For $\epsilon = 0$, $\lambda(l)$ tends to zero as $1/l$ and $\nu(l)$ has logarithmic corrections in l. For $\epsilon > 0$, the fixed point λ_1^* becomes unstable and two new stable fixed points,

$$\lambda_{2,3}^* = \pm\sqrt{\frac{\epsilon}{3A(d)}} , \qquad (4.5.54)$$

emerge. Here, the negative fixed point can be excluded by demanding the positivity of the coupling parameter. The stable fixed point $\lambda_2^* = \sqrt{\frac{\epsilon}{3A(d)}}$ leads to a renormalized viscosity that is independent of l provided that

$$z(l) = 2 - \epsilon/3 , \qquad (4.5.55)$$

which fixes the remaining parameter $a(l) = \int dl z(l) = (2 - \epsilon/3)l$. Moreover, the existence of a non-trivial fixed point λ_2^* strongly depends on the power law of the correlation function of the stirring force. A lower bound can be established from Eqs. (4.5.54) and (4.5.39) according to $y > d - 4$. The upper bound can be established via an argument that justifies the neglection of triple moments of $\mathbf{u}^<$ in Eq. (4.5.35) and yields $y < d$. The power law is thus bounded by

$$d - 4 < y < d. \tag{4.5.56}$$

The theory by Forster et al. [30] further predicts an energy spectrum of the form

$$E(k) = k^{-5/3 + 2(d-y)/3} \, . \tag{4.5.57}$$

This result coincides with Kolmogorov's prediction for $d = y$, i.e., if the power law of the spectrum y is equal to 3 for three-dimensional turbulence. It strictly applies for $k \to 0$ and generalizations to higher k are rather difficult, since they invalidate the low-order perturbation theory. Hence, the renormalization group once more has to be considered as a more informal treatment that has rather limited success in turbulence theory.

4.6 Chapter Conclusions

The emphasis of this chapter lied on the presentation of different closure methods. We have shown that closures which are based on a perturbative treatment of the nonlinearity fail to account for the intermittent character of the velocity field in turbulence.

Perhaps one of the most interesting quotes on the failing of renormalization and renormalization group methods in turbulence is by Migdal [32]:

> After trying for few years to do something with the Wyld approach I conclude that this is a dead end. The best bet here would be the renormalization group, which magically works in statistical physics. Those critical phenomena were close to Gaussian.
>
> The observed variety of vorticity structures with their long range interactions does not look like the block spins of critical phenomena. There is no such luck in turbulence. The nonlinear effects are much stronger[...]
>
> No! These old tricks are not going to work, we have to invent the new ones.

The above reasoning calls for *non-perturbative treatment* of the nonlinearity in the Navier-Stokes equation. In the following chapter, we will present such non-perturbative methods and try to assess their practicability within a comprehensive statistical description of turbulence.

References

1. Heisenberg, W.: Zur statistischen Theorie der Turbulenz. Zeitschrift für Phys. **124**(7), 628–657 (1948)
2. v. Weizsäcker, C.F.: Das Spektrum der Turbulenz bei großen Reynoldsschen Zahlen. Zeitschrift für Phys. **124**(7), 614–627 (1948)
3. Orszag, S.A.: On the elimination of aliasing in finite-difference schemes by filtering high-wavenumber components. J. Atmos. Sci. **28**(6), 1074 (1971)
4. Kraichnan, R.H.: Relation of fourth-order to second-order moments in stationary isotropic turbulence. Phys. Rev. **107**(6), 1485–1490 (1957)
5. Wyld, H.W.: Formulation of the theory of turbulence in an incompressible fluid. Ann. Phys. (N. Y) **14**, 143–165 (1961)
6. Kovasznay, L.S.: Spectrum of locally isotropic turbulence. J. Aeronaut. Sci. **15**(12), 745–753 (1948)
7. Monin, A.S., Yaglom, A.M.: Statistical Fluid Mechanics: Mechanics of Turbulence. Courier Dover Publications (2007)
8. Lesieur, M.: Turbulence in Fluids (2012)
9. Faust, G., Argyris, J., Haase, M., Friedrich, R.: An Exploration of Dynamical Systems and Chaos. Springer (2015)
10. Oboukhov, A.M.: Spectral energy distribution in a turbulent flow. Dokl. Akad. Nauk SSSR **1**(32), 22–24 (1941)
11. Kolmogorov, A.N.: The local structure of turbulence in incompressible viscous fluid for very large Reynolds numbers. Dokl. Akad. Nauk SSSR **30**(1890), 301–305 (1941)
12. Millionschikov, M.: On the theory of homogeneous and isotropic turbulence, p. 32. Dokl. Akad, Nauk SSSR (1941)
13. Ogura, Y.: A consequence of the zero-fourth-cumulant approximation in the decay of isotropic turbulence. J. Fluid Mech. **16**(1), 33–40 (1963)
14. Orszag, S.A.: Analytical theories of turbulence. J. Fluid Mech. **41**(2), 363–386 (1970)
15. McComb, W.D.: The Physics of Fluid Turbulence. Oxford University Press (1990)
16. Frisch, U., Lesieur, M., Brissaud, A.: A Markovian random coupling model for turbulence. J. Fluid Mech. **65**(01), 145–152 (1974)
17. Orszag, S.A.: Lectures on the statistical theory of turbulence. Flow Research (1974). Incorporated
18. Itzykson, C., Zuber, J.-B.: Quantum Field Theory. Dover (1980)
19. Carroll, S.: How Quantum Field Theory Becomes "Effective" (2013)
20. Huang, K.: Fundamental Forces of Nature: The Story of Gauge Fields. World Scientific Publishing Co Pte Ltd, Singapore (2007)
21. Haken, H., Wolf, H.C.: The Physics of Atoms and Quanta: Introduction to Experiments and Theory. Springer, Berlin Heidelberg (2004)
22. Ward, J.C.: An identity in quantum electrodynamics. Phys. Rev. **78**(2), 182 (1950)
23. Dyson, F.J.: Divergence of perturbation theory in quantum electrodynamics. Phys. Rev. **85**(4), 631–632 (1952)
24. Chandrasekhar, S.: The fluctuations of density in isotropic turbulence. Proc. R. Soc. London. Ser. A. Math. Phys. Sci. **210**(1100), 18–25 (1951)
25. McComb, W.D.: Renormalization Methods: A Guide For Beginners. Oxford University Press (2008)
26. Wilson, K.G.: The renormalization group: Critical phenomena and the Kondo problem. Rev. Mod. Phys. **47**(4), 773–840 (1975)
27. Onsager, L.: Crystal statistics. I. A two-dimensional model with an order-disorder transition. Phys. Rev. **65**(3-4), 117–149 (1944)
28. Landau, L.D., Lifshitz, E.M.: Statistical Physics, Third Edition: Volume 5 (Course of Theoretical Physics). Butterworth-Heinemann (1987)
29. Kadanoff, L.P.: Scaling laws for ising models near T_c. Phys. (Coll. Park. Md). **2**(233) (1966)

30. Forster, D., Nelson, D.R., Stephen, M.J.: Large-distance and long-time properties of a randomly stirred fluid. Phys. Rev. A **16**(2), 732–749 (1977)
31. Yakhot, V., Orszag, S.A.: Renormalization group analysis of turbulence. I. Basic theory. J. Sci. Comput. **1**(1), 3–51 (1986)
32. Migdal, A.A.: Turbulence as statistics of vortex cells. (1993) arXiv:hep-th/9306152

Chapter 5
Non-Perturbative Methods

The previous chapter highlighted the limitations of perturbative treatments of the Navier-Stokes equation such as the quasi-normal approximation and the renormalization or renormalization group method. It was found that nonlinearities are too strong to be grasped in a perturbative sense as perturbation expansions are set up in terms of powers of the Reynolds number. Therefore, the classical field theory of turbulence possesses no small parameter which necessarily demands a *non-perturbative* treatment of the Navier-Stokes equation.

In order to categorize the present chapter in more detail, we will briefly summarize the shortcomings of the closure hypothesis discussed in Chap. 4.

Reasons for the failing of the closure hypothesis of Chap. 4:

(i.) Due to high Reynolds numbers, fully developed turbulence does not exhibit a small parameter. Therefore, a perturbative treatment of the nonlinearity that starts from the purely linear regime such as the direct interaction approximation results in amiss predictions, e.g., an energy spectrum $E(k) \sim k^{-3/2}$ which violates Galilean invariance.

(ii.) Perturbation expansions are only valid for *widely separated* and *distinct* length and time scales. However, turbulent flows exhibit *no distinct separation of scales*.

(iii.) Moment closure schemes, such as the quasi-normal approximation, are based on properties of Gaussian-distributed velocity field fluctuations that are suitably extended to allow for non-vanishing moments of third order. Despite the fact that, by virtue of this extension, the quasi-normal approximation allows for an energy transfer across scales, it results in negative energy spectra that do *not* constitute a small effect (realizability).

(iv.) Velocity increments in a turbulent flow follow a strictly *non-self-similar* probability density function which is a direct implication of the effects of intermittency. Therefore, fluctuations on different scales are not *scale invariant*, which is a prerequisite for the applicability of the renormalization group.

© Springer Nature Switzerland AG 2021   105
J. Friedrich, *Non-perturbative Methods in Statistical Descriptions of Turbulence*,
Progress in Turbulence - Fundamentals and Applications 1,
https://doi.org/10.1007/978-3-030-51977-3_5

In the present chapter, we want to discuss three apparently different concepts that were developed over the past three decades of turbulence research: the operator product expansion, the instanton method, as well as a stochastic interpretation of the turbulent energy cascade in terms of Markov processes of velocity increments *in scale*. An inherent feature of each of the three methods is that they approximate the multi-point statistics described in Sect. 3 without relying upon any Gaussian or pseudo-Gaussian assumptions. Hence, such methods necessarily entail different physical arguments or apply only to some characteristic regimes. The so-called instanton method, for instance, solely applies to the far-tail behavior of certain probability density functions, e.g., instigated by extreme velocity gradient events. We will try to give an overview on how these different methods might be brought together in order to develop comprehensive non-perturbative treatments for the Navier-Stokes equation.

5.1 Fusion Rules and the Operator Product Expansion

Another approach to turbulence that emanated from quantum field theory is the so-called operator product expansion. In this approach, one mainly focuses on multi-scale velocity increment correlations such as

$$\langle v(x,r)^j v(x,r')^k \rangle = \langle (u(x+r) - u(x))^j (u(x+r') - u(x))^k \rangle \quad \text{for} \quad r \leq r'.$$
(5.1.1)

Here, we have narrowed our focus down to one-dimensional increments that possess the same point of reference x. The operator product expansion leads to so-called *fusion rules* that apply to this three-point correlation in the limit $r \to r'$ and leads to an expression that involves solely a product of structure functions $S_n(r) = \langle v(x,r)^n \rangle$, i.e., two-point quantities of certain orders. Before further discussing this procedure, it may be beneficial to shed some light on the historical development of fusion rules in general. The next section is therefore dedicated to fusion rules in the context of quantum chromodynamics, from which the entire procedure originated.

5.1.1 *Operator Product Expansion in Quantum Field Theory*

Originally, fusion rules were developed for the calculation of products of composite operators, known also as *local operators*, at short distance. The current algebra which enters the total hadronic annihilation cross section $\sigma_{e^+e^- \to hadrons}$ in quantum chromodynamics is a prominent example for such local operators [1] (for further discussions of current algebras, we also refer to [2]). Latter quantity can be derived in a first-order approximation in the electromagnetic interaction according to

$$\sigma_{e^+e^-\to hadrons} = \frac{e^4}{2(q^2)^3} L^{\mu\nu} \int d^4x e^{iq\cdot x} \langle 0|J^\mu(x)J^\nu(0)|0\rangle \,, \qquad (*5.1.2)$$

where $L^{\mu\nu} = [g^{\mu\nu}p_+ \cdot p_- - p_+^\mu p_+^\nu - p_+^\mu p_-^\nu]$ and $q^\mu = (p_+ + p_-)^\mu = q^0\delta_0^\mu$. Moreover, $p_\pm^\mu$ denotes the four momentum of the electron and positron, respectively. Furthermore, summation over equal Greek letters $0, 1, \ldots, 3$ is implied. The electromagnetic hadronic current is defined according to

$$J_\mu(x) = \sum_{quarks\ j} [e_j \bar{q}^j(x)\gamma_\mu q^j(x)] \,, \qquad (*5.1.3)$$

and enters in Eq. (*5.1.2) in form of a composite operator product vacuum expectation value. In the high energy limit, the main contribution to the integral in Eq. (*5.1.2) stems from the hadronic electromagnetic tensor in the limit of $x \to 0$, i.e.,

$$\lim_{x\to 0} \langle 0|J^\mu(x)J^\nu(0)|0\rangle \,. \qquad (*5.1.4)$$

This expression can be considered as the short distance behavior of the vacuum expectation value of the time-ordered product of two electromagnetic currents. In this first-order approximation, the total cross section (*5.1.2) is calculated as $\sigma_{e^+e^-\to hadrons} \approx \frac{4\pi\alpha^2}{3q^2} \sum_j e_j^2$, where α is the fine structure constant. However, this result is only of partial interest for the general problem: similar to hydrodynamic turbulence, quantum chromodynamics is a *strongly interacting system* and therefore not amenable to perturbation theory. Hence, computing the first-order approximation is not sufficient for the accurate determination of the general annihilation cross section which involves interactions in form of the general hadronic electromagnetic tensor $\langle J^\mu(x)J^\nu(0)\rangle$. Nevertheless, in the high energy limit, i.e., for, $\lim_{x\to 0} \langle J^\mu(x)J^\nu(0)\rangle$ Wilson [3] originally applied the operator product expansion and derived corresponding fusion rules. To put it more generally, we will discuss the operator product expansion at the example of the product of two operator valued fields $\hat{A}$ and $\hat{B}$, namely, $\hat{A}(x)\hat{B}(y)$. The basic idea now is that this bi-local operator can be expanded (for the case that y is near x) in terms of *local* operators C_j according to

$$\hat{A}(x)\hat{B}(y) = \sum_{j=1}^{n} \chi_j(x-y)\hat{C}_j(x) \,, \qquad (*5.1.5)$$

with the coefficient functions $\chi_j(x-y)$ that may be singular if $y = x$. Thus, the two key features of this operator product expansion are

(a) The two-point operator product in $A(x)B(y)$ was *fused* into a sum of one-point operators $C_j(x)$ and the corresponding correlation functions $\chi_j(x-y)$.
(b) The operator product expansion is a *non-perturbative* treatment of the interaction term and is therefore suitable for *strongly interacting physical systems* that are not amenable to common perturbative treatments.

In the following section, we will discuss fusion rules in the realm of turbulence theory.

5.1.2 Fusion Rules in the Statistical Theory of Turbulence

It is not quite obvious how aforementioned concepts from quantum field theory of strongly interacting systems were transferred to the statistical theory of turbulence. One of the first works that one has to mention is the multifractal approach by Paladin and Vulpiani [4]. Their concern was the scaling of the number of degrees of freedom N with the Reynolds number Re. They concluded that a satisfactory description of turbulence requires resolving scales up to a scale where molecular friction is of the same magnitude as the nonlinear energy transfer. In the framework of the K41 theory, this scale is the Kolmogorov dissipation length $\eta = (\nu^3/\langle\varepsilon\rangle)^{1/4}$. The Reynolds number can be cast as the non-dimensional ratio $\mathrm{Re} = (\langle\varepsilon\rangle L^4)^{1/3}/\nu$, where L is the integral length scale. Starting from the integral length scale L, the number of grid points per volume that allows for an accurate description of a three-dimensional fluid is

$$N(\mathrm{Re}) \sim \left(\frac{L}{\eta}\right)^3 \sim \mathrm{Re}^{9/4} . \tag{5.1.6}$$

This result was first discussed by Landau and Lifshitz [5] and has already been mentioned in Chap. 2, telling us that the number of degrees of freedom is an increasing function of the Reynolds number. The central assumption that led to Eq. (5.1.6) was that the local energy dissipation rate $\varepsilon(\mathbf{x})$ is smoothly distributed on a homogeneous three-dimensional region. A generalization of this result can be obtained through assuming that the local energy dissipation rate $\varepsilon(\mathbf{x})$ is concentrated on a homogeneous fractal with (non-integer) Hausdorff dimension $D_f < 3$. The latter assumption corresponds to the so-called β-model. The longitudinal velocity increments in this approach are supposed to scale as

$$v(\mathbf{x}, r) = (\mathbf{u}(\mathbf{x} + \mathbf{r}) - \mathbf{u}(\mathbf{x})) \cdot \frac{\mathbf{r}}{r} \sim r^h , \tag{5.1.7}$$

where $h = (D_f - 2)/3$. The dissipation length η can now be retrieved by imposing that

$$\mathrm{Re}_\eta = \frac{v(\mathbf{x}, \eta)\eta}{\nu} \sim \mathcal{O}(1) , \tag{5.1.8}$$

which yields $\eta \sim L/\mathrm{Re}^{1/1+h}$. Note that for the K41 phenomenology $h = 1/3$, this yields the Kolmogorov dissipation length from above. The number of degrees of freedom can be estimated as

$$N(\mathrm{Re}) \sim \left(\frac{L}{\eta}\right)^{D_f} \sim \mathrm{Re}^{3D_f/(1+D_f)} . \tag{5.1.9}$$

As it has been mentioned in Sect. 3.2.3, the assumption of a homogeneous distribution of the local energy dissipation rate is not fulfilled in three-dimensional turbulence due to the presence of intermittency effects, which are in contradiction with scaling exponents of the homogenous fractal model, $\zeta_n = \frac{D_f - 2}{3} n + 3 - D_f$. Further progress can be made, however, by assuming that $\varepsilon(\mathbf{x})$ is located on an inhomogeneous fractal.

In the so-called multifractal model [6, 7], local scale invariance is implied by the existence of a continuous range of scaling exponents h: For any h in this range, there is a set $S(h)$ in $\mathbb{R}^3$ of Hausdorff dimension $D(h)$ such that when $\mathbf{r} \in S(h)$ the velocity increment for small r is given according to $v(\mathbf{x}, r) \sim r^h$. Now each set contributes to the structure function of order n with a probability of encountering the set $S(h)$ within a ball of radius r that is controlled by the co-dimension $3 - D(h)$ according to $\sim r^{3-D(h)}$. Therefore, the structure functions of the multifractal approach can be obtained according to

$$S_n(r) = \langle v(\mathbf{x}, r)^n \rangle = \int d\mu(r) r^{nh+3-D(h)} , \qquad (5.1.10)$$

where $\mu(r)$ is a suitable weighting factor. We can finalize these calculations via a saddle point approximation of the integral in the limit of small r, which yields

$$S_n(r) \sim r^{\zeta_n} , \qquad \text{where} \qquad \zeta_n = \min_h[nh + 3 - D(h)] . \qquad (5.1.11)$$

The scaling exponent ζ_n can thus be obtained from a Legendre transformation of the co-dimension. The knowledge of the number of degrees of freedom that is necessary for describing the multifractal cascade, however, is far from obvious. Perhaps the most interesting point which was made by Paladin and Vulpiani is that for each singularity h a different $\eta(h)$ is obtained from the condition $\eta \sim \mathrm{Re}^{-1/1+h}$. Therefore, a continuous range of dissipation lengths is excited and the corresponding number of degrees of freedom has to be acquired via an integration over h. A similar argument was put forward by Nelkin who discussed a fluctuating Kolmogorov scale $\eta(h)$ in the context of the multifractal picture [8]. The total number of degrees of freedom for the multifractal model reads

$$N(\mathrm{Re}) \sim \int d\mu(h) \left(\frac{L}{\eta(h)} \right)^{D(h)} \sim \int d\mu(h) \mathrm{Re}^{D(h)/(1+h)} \sim \mathrm{Re}^\delta , \qquad (5.1.12)$$

where $\delta = \max_h[D(h)/(1 + h)]$. We can get the exponent δ from a fit of experimental data, for instance, the data by Anselmet et al. [9] which suggests $\delta = 2.2$. This result is actually not so far from Landau's estimate (5.1.6). It is also interesting to determine the minimal scaling exponent $h_{\min}$ from experiments which show that $h_{\min} < 0.08$. The log-normal model by Kolmogorov [10] and Oboukhov [11] exhibits scaling with negative exponents which implies an unbounded velocity. Here, the velocity becomes greater than the speed of sound, necessarily violating the incompressibility condition (2.1.2). Another implication of the h-dependent viscous cutoff in the multifractal model is the existence of a so-called *intermediate dissipation range* put forth in [7]. The integration in Eq. (5.1.10) is thus limited to those scaling exponents h that guarantee a viscous cutoff $\eta(h) < r$. As an example, we consider the structure function of second order

$$S_2(r) \sim \int_{\eta(h)<r} d\mu(h) r^{2h+3-D(h)} . \tag{5.1.13}$$

If we further define h_K to be the dominant scaling exponent in the integration and respect that deviations from K41 scaling are rather small, we can estimate $h_K \approx 1/3$. Obviously, this corresponds to the viscous cutoff $\eta(h_K) \sim \nu^{3/4}$. In the inertial range $\eta(h_K) \ll r \ll L$, we obtain the usual scaling $\zeta_2 = 2h_K + 3 - D(h_K) \approx 2/3$. In the intermediate dissipation range $\eta(h_{min}) < r < \eta(h_K)$, however, we retrieve a new scaling that is due to the suppression of contributions from h_K in favor of lesser scaling exponents that are not yet influenced by viscosity. Furthermore, in the limit of small ν, the dominant contributions to the integral in Eq. (5.1.13) stem from the scaling exponent $h(r)$ that belongs to $\eta(h) = r$, which yields

$$S_2(r) \sim r^{2h(r)+3-D(h)} , \qquad \nu^{1/(1+h(r))} = r , \tag{5.1.14}$$

for $\eta(h_{min}) < r < \eta(h_K)$. This translates to an energy spectrum of the form

$$E(k) \sim k^{-1-2h(k)-[3-D(h(k))]} , \quad \text{with} \quad h(k) = -1 - \frac{\log \nu}{\log k} , \tag{5.1.15}$$

which is a power law with slowly varying exponent, also referred to as pseudoalgebraic law. The multi-scaling approach [7] is of particular interest with respect to multi-scale velocity increment correlations (5.1.1) in the dissipation range. In this case, the velocity increment approaches the velocity gradient and one is left with a joint velocity increment-velocity gradient correlation, e.g.,

$$\left\langle \left(\frac{\partial u(x)}{\partial x} \right)^2 v(x, r')^k \right\rangle , \tag{5.1.16}$$

which can be derived for $j = 2$ in Eq. (5.1.1). This correlation can be recast in assuming that r in Eq. (5.1.1) is located in the intermediate dissipation, i.e., $\eta(h_{min}) < r < \eta(h_K)$, resulting in

$$\left\langle \left(\frac{\partial u(x)}{\partial x} \right)^2 v(x, r')^k \right\rangle \sim \left\langle \left(\frac{v(x, r)}{r} \right)^2 v(x, r')^k \right\rangle , \tag{5.1.17}$$

which was proposed in [12, 13]. As in the multifractal model, some other manipulations of this formula are : first of all, it is assumed that the small-scale increment $v(x, r)$ is connected to the inertial range increment $v(x, r')$ via the multiplier $w(r, r')$

according to

$$v(x, r) \sim w(r, r')v(x, r') \quad \text{with} \quad w(r, r') = \left(\frac{r}{r'}\right)^h , \tag{5.1.18}$$

with the probability $(r/r')^{3-D(h)}$. Furthermore, the most dominant contribution to the correlation (5.1.17) emerges from the scaling exponent that belongs to $\eta(h) = r$, which implies

$$v(x, r)r \sim \left(\frac{r}{r'}\right)^h v(x, r')r \sim \nu . \tag{5.1.19}$$

Similar to the second-order structure function with viscous cutoff (5.1.13), the correlation (5.1.17) can be cast as

$$\left\langle \left(\frac{\partial u(x)}{\partial x}\right)^2 v(x, r')^k \right\rangle$$
$$\sim \int_{\eta(h)<r} d\mu(h) \frac{v(x, r')^{k+2}}{r'^2} \left(\frac{\nu}{r'v(x, r')}\right)^{[2(h-1)+3-D(h)]/(1+h)} . \tag{5.1.20}$$

For the special case $k = 0$, we know that

$$\left\langle \left(\frac{\partial u(x)}{\partial x}\right)^2 \right\rangle \sim \nu^{-1} . \tag{5.1.21}$$

Therefore, Benzi et al. [12, 13] deduced that

$$\left\langle \left(\frac{\partial u(x)}{\partial x}\right)^2 v(x, r')^k \right\rangle \sim \frac{r'^{\zeta_{k+3}}}{\nu r'} , \tag{5.1.22}$$

in the limit of small ν. Hence, this relation represents a dissipation anomaly: The quantity $\nu \left\langle \left(\frac{\partial u(x)}{\partial x}\right)^2 v(x, r')^k \right\rangle$ remains finite in the limit $\nu \to 0$, similarly to the discussion at the example of shocks in Burgers equation in Sect. 2.4. As a matter of fact, the multifractal prediction (5.1.22) can be derived directly from the Burgers equation which will be discussed in Sect. 5.4.1. However, at this point, it is absolutely unclear whether this holds true for the general case of the Navier-Stokes equation where pressure contributions have to be considered as well.

We are now able to generalize this result to an arbitrary power of the velocity gradient as well. To this end, we consider the generalization of Eq. (5.1.20)

$$\left\langle \left(\frac{\partial u(x)}{\partial x} \right)^j v(x,r')^k \right\rangle$$

$$\sim \int_{\eta(h)<r} \mathrm{d}\mu(h) \frac{v(x,r')^{k+2}}{r'^j} \left(\frac{\nu}{r'v(x,r')} \right)^{[j(h-1)+3-D(h)]/(1+h)} . \quad (5.1.23)$$

Under the assumption that the velocity gradient possesses moments that scale in terms of the viscosity according to

$$\left\langle \left(\frac{\partial u(x)}{\partial x} \right)^j \right\rangle \sim \frac{1}{\nu^{\xi_j}} , \quad (5.1.24)$$

we obtain for $\nu \to 0$

$$\left\langle \left(\frac{\partial u(x)}{\partial x} \right)^j v(x,r')^k \right\rangle \sim \frac{r'^{\zeta_{k+j}+\xi_j} r'^{j-\xi_j}}{\nu^{\xi_j}} . \quad (5.1.25)$$

Here, the scaling exponents of the velocity gradient and that of the structure function are related by $\xi_j = (l - \zeta_l)/2$ and $j = (\zeta_l + l)/2$ (see [6]). For instance, for $j = 2$, we obtain $\xi_2 = 1$ under the condition that $\zeta_3 = 1$, which reduces Eqs. (5.1.25) to (5.1.22).

In the last part of this section, we will discuss the fusion rules which can be applied to the multi-scale correlations (5.1.1) for the case when *both* r and r' lie within the inertial range. The central assumption is again that small-scale statistics are related to the large-scale configuration via the multiplier $w(r, r')$ according to Eq. (5.1.18). Both anomalous scaling and viscous effects can be reproduced by choosing a suitable random process for the multiplier $w(r, r')$. In particular, a purely uncorrelated multiplicative process in addition to homogeneity along the energy cascade suggests that

$$w(r,r') \to w(r/r') . \quad (5.1.26)$$

Inserting the relation (5.1.18) into Eq. (5.1.1) yields

$$\langle v(x,r)^j v(x,r')^k \rangle \sim \langle w(r/r')^j v(x,r')^{j+k} \rangle \sim \langle w(r/r')^j w(r'/L)^{j+k} \rangle S_{j+k}(L) ,$$

where we further assumed that the large-scale configuration is related to the integral scale increment by the same relation (5.1.18), which then enabled us to pull the structure function at the integral length scale out of the averages. Next, assuming that the multiplicative process is uncorrelated, we can factorize the average $\langle w(r/r')^j w(r'/L)^{j+k} \rangle = \langle w(r/r')^j \rangle \langle w(r'/L)^{j+k} \rangle$ and make use of $S_j(r) = \langle w(r/L) \rangle S_j(L)$. The fusion rules for the multi-scale velocity increment thus read

$$\langle v(x,r)^j v(x,r')^k \rangle \sim \frac{S_j(r)}{S_j(r')} S_{j+k}(r') \quad \text{for} \quad r < r' . \tag{5.1.27}$$

These relations have shown to be in quantitative agreement with experimental multi-scale analysis [12, 13] for small-scale separations r/r'. Equation (5.1.27) was first discussed by Eyink [14] and subsequently apprehended by L'vov and Procaccia [15]. In analogy to the operator product expansion from field theory, the fusion rules reduce the three-point quantity to terms that only involve two-point quantities and, apparently, do not require an explicit scaling behavior of structure functions. It has been suggested by Benzi et al. [13] that this implies an independence of the fusion rules from the Reynolds number. Nevertheless, the validity of fusion rules is questionable for $r' \to r$. In this limit, correlations between the two increments are the strongest and involve structure functions that are located in the dissipation range, which are not accounted for by the fusion rules (5.1.27). This is apparent from the following identity:

$$S_n(r' - r) = \langle (u(x + r' - r) - u(x))^n \rangle = \langle (u(x + r') - u(x + r))^n \rangle . \tag{5.1.28}$$

Here we relied on the assumption of homogeneity and shifted the point of reference x to $x + r$.

A further treatment of the terms leads to

$$S_n(r' - r) = \langle (u(x + r') - u(x) + u(x) - u(x + r))^n \rangle = \langle (v(x,r') - v(x,r))^n \rangle , \tag{5.1.29}$$

or

$$S_n(r' - r) = \sum_k^n \binom{n}{k} (-1)^k \langle v(x,r')^k v(x,r)^{n-k} \rangle . \tag{5.1.30}$$

This exact identity thus relates structure functions of order n to multi-scale velocity increment correlations, which is a direct consequence of the assumption of homogeneity. For $n = 2$, for instance, this yields

$$S_2(r' - r) = S_2(r') - 2\langle v(x,r')v(x,r) \rangle + S_2(r) , \tag{5.1.31}$$

and for $n = 3$, we obtain

$$S_3(r' - r) = S_3(r') - 3\langle v(x,r')^2 v(x,r) \rangle + 3\langle v(x,r')v(x,r)^2 \rangle - S_3(r) . \tag{5.1.32}$$

Obviously, these exact relations differ from the fusion rules (5.1.27) in the limit $r' \to r$, since fusion rules do not involve small-scale quantities such as $S_n(r' - r)$. Another shortcoming can be deduced for the case $j = k = 1$. In this case, the fusion rules (5.1.27) reduce to

$$\langle v(x, r')v(x, r) \rangle = \frac{S_1(r)}{S_1(r')} S_2(r') , \qquad (5.1.33)$$

which is undetermined, since $S_1(r) = \langle v(x, r) \rangle = 0$, due to homogeneity. On the other hand, we obtain from Eq. (5.1.31) the following relation:

$$\lim_{r' \to r} \frac{\partial^2 \langle v(x, r')v(x, r) \rangle}{\partial r'^2} = \frac{-S_2''(0) + S_2''(r)}{2} = -\frac{\langle \varepsilon \rangle}{2\nu} + \frac{S_2''(r)}{2} , \qquad (5.1.34)$$

which is clearly different from zero. This relation is of particular interest for the multi-point PDF hierarchy discussed in Sect. 5.4.

5.2 Stochastic Interpretation of the Turbulent Energy Cascade

The purpose of this section is to discuss the phenomenological model by R. Friedrich and J. Peinke [16] (see also [17] for a comprehensive review) that interprets the turbulent energy cascade as a Markov process of velocity increments *in scale*. The vigor of the Friedrich-Peinke phenomenology is that it allows for an effective description of the phenomenon of intermittency. Since the Markov property leads to a considerable reduction of degrees of freedom *and* preserves the intermittent character of velocity field fluctuations, it can be considered as an appropriate closure method for the hierarchies of turbulence discussed in Chap. 3. We will show explicitly that the phenomenology is capable to reproduce the essence of anomalous scaling. Therefore, it also allows for an adequate interpretation of intermittency in terms of the evolution of the velocity increment PDF in scale via a so-called Kramers-Moyal expansion.

5.2.1 Markov Process of Velocity Increments in Scale

In their seminal paper [16], Friedrich and Peinke investigated the multi-scale velocity statistics in a free jet experiment. The key quantity in their approach is the n-increment PDF of the longitudinal velocity increments

$$v(\mathbf{r}, \mathbf{x}, t) = (\mathbf{u}(\mathbf{x} + \mathbf{r}, t) - \mathbf{u}(\mathbf{x}, t)) \cdot \frac{\mathbf{r}}{r} , \qquad (5.2.1)$$

which is defined according to

$$f_n(v_n, r_n; \ldots; v_1, r_1, \mathbf{x}, t) = \prod_{i=1}^{n} \langle \delta(v_i - v(r_i, \mathbf{x}, t)) \rangle . \qquad (5.2.2)$$

The n-increment PDF is a high-dimensional object whose determination from first principles is inaccessible due to the hierarchical ordering that is inherent in a turbulent flow (see Sect. 3.4 for further discussion). In the following, we will focus on the spatial properties of the n-increment PDF at different scales r_i, i.e., we will assume stationarity. Due to the left-bounded velocity increment definition (5.2.1), we can further assume homogeneity with respect to the point of reference $\mathbf{x}$.

The theory of stochastic processes [18] suggests that the n-increment PDF can be expressed as a product of the $n - 1$-increment PDF and a conditional probability

$$p(v_n, r_n | v_{n-1}, r_{n-1}; \ldots; v_1, r_1) = \frac{f_n(v_n, r_n; \ldots; v_1, r_1)}{f_{n-1}(v_{n-1}, r_{n-1}; \ldots; v_1, r_1)} . \qquad (5.2.3)$$

The conditional probability density functions can now be put into the context of the turbulent energy cascade in the following way: as it has been discussed in Sect. 3.2, the central notion of the energy cascade is the transport of energy from eddies of larger scales to eddies of smaller scales (for the notions of eddies see Sect. 2.3.2) where the energy is ultimately transformed into heat. Here, the assumption of a hierarchical organization of eddies implies that energy is solely transferred between eddies of adjacent scales. In applying this concept to the multi-scale approach that we have undertaken via Eqs. (5.2.2) and (5.2.3), we can interpret the velocity field v_i at a scale r_i to be associated with an eddy of size r_i that possesses the kinetic energy $\frac{1}{2}v_i^2$. Considering three eddies that belong to v_3, v_2 and v_1 with $r_3 \leq r_2 \leq r_1$, the hierarchical organization that is depicted in Fig. 5.1 implies that the conditional probability satisfies

$$p(v_3, r_3 | v_2, r_2; v_1, r_1) = p(v_3, r_3 | v_2, r_2) \quad \text{for} \quad r_3 \leq r_2 \leq r_1 , \quad (5.2.4)$$

which is a Markov property of the velocity increments in scale. For the general case of the $n - 1$-times conditional PDF, the Markov property implies

$$p(v_n, r_n | v_{n-1}, r_{n-1}; \ldots; v_1, r_1) = p(v_n, r_n | v_{n-1}, r_{n-1}) . \qquad (5.2.5)$$

Consequently, the Markov property leads to a tremendous reduction of the complexity of the problem: the whole statistics of the turbulent cascade is now determined by the transition probabilities $p(v_i, r_i | v_{i-1}, r_{i-1})$. This can be seen from the n-increment PDF, which can be rewritten as a chain of transition probabilities

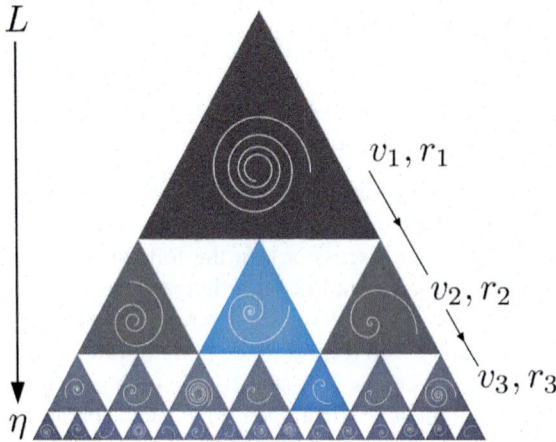

Fig. 5.1 Schematic depiction of the concept of the turbulent energy cascade in the realm of the Friedrich-Peinke phenomenology. A velocity field fluctuation v_i at scale r_i can be associated with an eddy of size r_i that possesses the kinetic energy $\frac{1}{2}v_i^2$. The eddies of different sizes r_i are supposed to be ordered hierarchically in the inertial range, i.e., from the integral length L scale down to the dissipation scale η. In the following, we want to consider three particular eddies of size $r_3 \leq r_2 \leq r_1$ that belong to v_3, v_2, and v_1. Due to the localness of interactions required by the assumption of a hierarchical organization, the conditional probability $p(v_3, r_3|v_2, r_2; v_1, r_1)$ is independent upon the condition on v_1, r_1 and thus obeys the Markov property (5.2.4)

$$
\begin{aligned}
f_n(v_n, r_n; \ldots; v_1, r_1) = \; & p(v_n, r_n|v_{n-1}, r_{n-1}) \times p(v_{n-1}, r_{n-1}|v_{n-2}, r_{n-2}) \times \ldots \\
& \times p(v_2, r_2|v_1, r_1) \times f_1(v_1, r_1) .
\end{aligned} \tag{5.2.6}
$$

Therefore, under the presumption that $f_1(v_1, r_1)$ is known at large scales (typically a Gaussian distribution), the knowledge of the transition probability is sufficient for the determination of the n-increment PDF. The next section will discuss certain important consequences of the Markov property and its limitations.

5.2.1.1 Chapman-Kolmogorov Equation

An important equation for the transition PDFs can be derived from the Markov property and is known as the Chapman-Kolmogorov equation

$$
p(v_3, r_3|v_1, r_1) = \int dv_2\, p(v_3, r_3|v_2, r_2) p(v_2, r_2|v_1, r_1) \quad \text{for} \quad r_3 \leq r_2 \leq r_1 . \tag{5.2.7}
$$

The Chapman-Kolmogorov equation thus relates the transition PDFs of different pairs of scale separations. It is actually another statement of the reduction property of the joint increment PDFs,

$$f_{n-1}(v_{n-1}, r_{n-1}; \ldots; v_1, r_1) = \int dv_n\, f_n(v_n, r_n; \ldots; v_1, r_1)\,, \tag{5.2.8}$$

that are subject to the Markov property. Another important property that we want to state here for further use is the coincidence property of the conditional PDF (5.2.3)

$$\lim_{r_n \to r_{n-1}} p(v_n, r_n | v_{n-1}, r_{n-1}; \ldots; v_1, r_1) = \delta(v_n - v_{n-1})\,. \tag{5.2.9}$$

The derivation of the Chapman-Kolmogorov equation is straightforward. First of all, we consider the three-increment PDF $f_3(v_3, r_3; v_2, r_2; v_1, r_1)$ for $r_3 \le r_2 \le r_1$ and integrate over v_2

$$\int dv_2\, f_3(v_3, r_3; v_2, r_2; v_1, r_1) = f_2(v_3, r_3; v_1, r_1)\,, \tag{5.2.10}$$

where we made use of the reduction property (5.2.8). Both sides of this equation can now be written in terms of conditional probabilities according to

$$\int dv_2\, \underbrace{p(v_3, r_3 | v_2, r_2; v_1, r_1)}_{\text{Markov property}} p(v_2, r_2 | v_1, r_1) f_1(v_1, r_1) = p(v_3, r_3 | v_1, r_1) f_1(v_1, r_1)$$

$$= \int dv_2\, p(v_3, r_3 | v_2, r_2) p(v_2, r_2 | v_1, r_1) f_1(v_1, r_1) = p(v_3, r_3 | v_1, r_1) f_1(v_1, r_1)\,.$$

A final division by the one-increment PDF yields the Chapman-Kolmogorov equation (5.2.7).

5.2.1.2 The Markov-Einstein Length

The Markov property (5.2.4) has already been verified experimentally in free jet experiments [16, 19, 20] as well as numerically for two-dimensional turbulence [21]. Nevertheless, subsequent investigations [20, 22] showed that the Markov property is violated at small-scale separations $r_2 - r_3 \le \lambda_{ME}$. The prediction of such a limiting scale for a stochastic process was already noticed by Einstein in his famous paper on Brownian motion [23] and is therefore termed the Markov-Einstein length λ_{ME}. Einstein explicitly stated that two successive steps in a stochastic description of a Brownian diffusion process have to exceed the mean free path length, otherwise the two events become correlated and the stochastic description breaks down.

Interestingly, Lück, Peinke and Friedrich [22] could demonstrate that the Markov-Einstein length is of the order of the Taylor length discussed in Sect. 3.2.1.4. Since the

Taylor length λ is associated to the curvature of the smallest structures in a turbulent flow, e.g., the curvature of vorticity filaments in the three-dimensional Navier-Stokes equation, the Markov property is connected to the *roughness* of the velocity field. Consequently, the Markov property breaks down if scale separations $r_2 - r_3$ in Eq. (5.2.4) approach a scale, for which velocity field fluctuations become *smooth*.

From a mathematical point of view, the Markov-Einstein length corresponds to a scale beyond which the velocity field possesses continuous derivatives of all orders. Although experiments [22] suggest that the Markov-Einstein length λ_{ME} is related to the Taylor length, it is absolutely unclear whether this is a one-to-one correspondence. Moreover, the determination of λ_{ME} is far from obvious, since the breakdown of the Markov property happens in a more or less continuous fashion. Nevertheless, the violation of the Markov property underlines once more that a turbulent flow does not exhibit a separation of scales.

5.2.1.3 *Derivation of Generalized Kramers-Moyal Expansions

In this section, we want to derive evolution equations for PDFs and conditional PDFs in scale for a *generalized stochastic process* [24]. We will demonstrate that the latter can be greatly simplified under the assumption of a Markov process in scale yielding the so-called *Kramers-Moyal expansions* that are of great importance for relating the Markov property of velocity increments to phenomenological models of turbulence (see Sect. 5.2.2).

In using Bayes' theorem (5.2.3), the $n+1$-increment PDF can be cast according to

$$f_{n+1}(v, r; v_n, r_n; \ldots; v_1, r_1)$$
$$= p(v_n, r_n | v, r; \ldots; v_1, r_1) f_n(v, r; v_{n-1}, r_{n-1}; \ldots; v_1, r_1) . \quad (*5.2.11)$$

Integrating this equation with respect to v and making use of the reduction property (5.2.8) yields

$$f_n(v_n, r_n; \ldots; v_1, r_1)$$
$$= \int dv \, p(v_n, r_n | v, r; \ldots; v_1, r_1) f_n(v, r; v_{n-1}, r_{n-1}; \ldots; v_1, r_1) . \quad (*5.2.12)$$

In order to obtain an evolution equation for the n-increment PDF, we derive Eq. (*5.2.12) with respect to r_n and then fuse $r \to r_n$ on the r.h.s., which yields

$$\frac{\partial}{\partial r_n} f_n(v_n, r_n; \ldots; v_1, r_1)$$
$$= \lim_{r \to r_n} \int dv \, \frac{\partial p(v_n, r_n | v, r; \ldots; v_1, r_1)}{\partial r_n} f_n(v, r; v_{n-1}, r_{n-1}; \ldots; v_1, r_1) . \quad (*5.2.13)$$

The evolution in scale of the stochastic process is thus fully determined by the differential conditional PDF

$$
\lim_{r \to r_n} \frac{\partial p(v_n, r_n | v, r; \ldots; v_1, r_1)}{\partial r_n}
$$
$$
= - \lim_{\lambda \to 0} \frac{p(v_n, r_n - \lambda | v, r_n; \ldots; v_1, r_1) - \delta(v_n - v)}{\lambda} , \qquad (*5.2.14)
$$

where we made use of the coincidence property of the conditional PDF (5.2.9) and respected the ordering $r_n \leq r \leq r_{n-1} \leq \ldots \leq r_1$.

Let us further consider the Fourier transform

$$
\hat{p}(u, r_n | v, r; \ldots; v_1, r_1) = \int dv_n \, e^{iu(v_n - v)} p(v_n, r_n | v, r; \ldots; v_1, r_1) . \quad (*5.2.15)
$$

Expanding the exponential function in a power series in u yields

$$
\hat{p}(u, r_n | v, r; \ldots; v_1, r_1) = \sum_{k=0}^{\infty} (iu)^k M^{(k)}(v, r; r_n | v_{n-1}, r_{n-1}; \ldots; v_1, r_1) ,
$$
$$
(*5.2.16)
$$

where

$$
M^{(k)}(v, r; r_n | v_{n-1}, r_{n-1}; \ldots; v_1, r_1)
$$
$$
= \frac{1}{k!} \int dv_n \, (v_n - v)^k p(v_n, r_n | v, r; \ldots; v_1, r_1) \qquad (*5.2.17)
$$

are defined as the conditional moments. Taking the inverse Fourier transform of Eq. (*5.2.16) yields

$$
p(v_n, r_n | v, r; \ldots; v_1, r_1)
$$
$$
= \frac{1}{2\pi} \sum_{k=0}^{\infty} \int du \, e^{-iu(v_n - v)} (iu)^k M^{(k)}(v, r; r_n | v_{n-1}, r_{n-1}; \ldots; v_1, r_1)
$$
$$
= \sum_{k=0}^{\infty} M^{(k)}(v, r; r_n | v_{n-1}, r_{n-1}; \ldots; v_1, r_1) \frac{\partial^k \delta(v_n - v)}{\partial v^k} , \qquad (*5.2.18)
$$

where we made use of the Fourier representation of the delta function according to

$$
\frac{1}{2\pi} \int du (iu)^k e^{-iu(v_n - v)} = \frac{\partial^k \delta(v_n - v)}{\partial v^k} . \qquad (*5.2.19)
$$

Inserting the expression for the conditional PDF (*5.2.18) into the differential conditional PDF (*5.2.14) and then into Eq. (*5.2.13) yields

$$\frac{\partial}{\partial r_n} f_n(v_n, r_n; \ldots; v_1, r_1) = -\sum_{k=1}^{\infty} \int dv \, D^{(k)}(v, r_n|v_{n-1}, r_{n-1}; \ldots; v_1, r_1)$$

$$\times \frac{\partial^k \delta(v_n - v)}{\partial v^k} f_n(v, r; v_{n-1}, r_{n-1}; \ldots; v_1, r_1), \qquad (*5.2.20)$$

where we want to term the coefficients

$$D^{(k)}(v, r_n|v_{n-1}, r_{n-1}; \ldots; v_1, r_1)$$

$$= \lim_{\lambda \to 0} \frac{M^{(k)}(v, r_n; r_n - \lambda|v_{n-1}, r_{n-1}; \ldots; v_1, r_1)}{\lambda}$$

$$= \frac{1}{k!} \lim_{\lambda \to 0} \frac{1}{\lambda} \int dv_n \, (v_n - v)^k p(v_n, r_n - \lambda|v, r_n; \ldots; v_1, r_1) \,, \quad (*5.2.21)$$

as the *generalized Kramers-Moyal coefficients*. Note that the $k = 0$-term in the sum dropped out due to the identity

$$M^{(0)}(v, r; r_n|v_{n-1}, r_{n-1}; \ldots; v_1, r_1) = \int dv_n \, p(v_n, r_n|v, r; \ldots; v_1, r_1) = 1 \,.$$

Partial integrations (k-times) of each term in the sum of Eq. (*5.2.20)

$$-\frac{\partial}{\partial r_n} f_n(v_n, r_n; \ldots; v_1, r_1) \qquad (*5.2.22)$$

$$= \sum_{k=1}^{\infty} \left(-\frac{\partial}{\partial v_n} \right)^k D^{(k)}(v_n, r_n|v_{n-1}, r_{n-1}; \ldots; v_1, r_1) f_n(v_n, r_n; \ldots; v_1, r_1) \,.$$

In dividing this equation by the $n - 1$-increment PDF it becomes apparent that the same evolution equation also holds for the conditional PDF

$$-\frac{\partial}{\partial r_n} p(v_n, r_n|v_{n-1}, r_{n-1}; \ldots; v_1, r_1) \qquad (*5.2.23)$$

$$= \sum_{k=1}^{\infty} \left(-\frac{\partial}{\partial v_n} \right)^k D^{(k)}(v_n, r_n|v_{n-1}, r_{n-1}; \ldots; v_1, r_1) p(v_n, r_n|v_{n-1}, r_{n-1}; \ldots; v_1, r_1) \,.$$

It must be stressed that Eq. (*5.2.23) reveals a hierarchical structure, i.e., the evolution in scale of the n-increment conditional PDF $p(v_n, r_n|v_{n-1}, r_{n-1}; \ldots; v_1, r_1)$ depends on the $n + 1$-increment conditional PDF $p(v_{n+1}, r_{n+1}|v_n, r_n; \ldots; v_1, r_1)$ that enters through the generalized Kramers-Moyal coefficients (*5.2.21). If the stochastic process is a Markov process in scale, however, the hierarchy ends at the evolution equation of the transition PDF

$$-\frac{\partial}{\partial r_2} p(v_2, r_2|v_1, r_1) = \sum_{k=1}^{\infty} \left(-\frac{\partial}{\partial v_2}\right)^k \underbrace{D^{(k)}(v_2, r_2|v_1, r_1)}_{\text{Markov: } D^{(k)}(v_2, r_2)} p(v_2, r_2|v_1, r_1),$$

(5.2.24)

which can be derived from the generalized Kramers-Moyal coefficient

$$D^{(k)}(v_2, r_2|v_1, r_1) = \frac{1}{k!} \lim_{\lambda \to 0} \frac{1}{\lambda} \int dv_3 \, (v_3 - v_2)^k \underbrace{p(v_3, r_2 - \lambda|v_2, r_2; v_1, r_1)}_{\text{Eq. (5.2.4)}}$$

$$= \frac{1}{k!} \lim_{\lambda \to 0} \frac{1}{\lambda} \int dv_3 \, (v_3 - v_2)^k p(v_3, r_2 - \lambda|v_2, r_2) = D^{(k)}(v_2, r_2).$$

Hence, the Markov property suggests that the transition PDF obeys the same evolution equation as the one-increment PDF, which will be discussed in the next section.

5.2.1.4 Kramers-Moyal Expansions

A central notion of a Markov process is that the one-increment PDF and the transition PDF follow the same Kramers-Moyal expansion in scale [18]

$$-\frac{\partial}{\partial r_1} f_1(v_1, r_1) = \sum_{k=1}^{\infty} \left(-\frac{\partial}{\partial v_1}\right)^k D^{(k)}(v_1, r_1) f_1(v_1, r_1),$$

(5.2.25)

$$-\frac{\partial}{\partial r_2} p(v_2, r_2|v_1, r_1) = \sum_{k=1}^{\infty} \left(-\frac{\partial}{\partial v_2}\right)^k D^{(k)}(v_2, r_2) p(v_2, r_2|v_1, r_1),$$

(5.2.26)

where the Kramers-Moyal coefficients are defined according to

$$D^{(k)}(v_1, r_1) = \frac{1}{k!} \lim_{r_2 \to r_1} \frac{1}{r_1 - r_2} \int dv_2 (v_2 - v_1)^k p(v_2, r_2|v_1, r_1).$$

(5.2.27)

Here, the minus signs in Eq. (5.2.25) and Eq. (5.2.26) indicate that the process runs from large to small scales. Moreover, Eq. (5.2.26) is a differential form of the Chapman-Kolmogorov equation (5.2.7) and is therefore only valid for the case of a Markov process [25] (see also Sect. 5.2.1.3 for the derivation of generalized Kramers-

Moyal expansions). Kramers-Moyal expansions allow for a suitable description of intermittency in terms of the evolution of the one-increment PDF in scale (5.2.25). The next section explicitly shows that Kramers-Moyal expansions of the Friedrich-Peinke approach are general enough to reproduce all currently known phenomenologies of intermittency in turbulence.

5.2.2   Relation of the Friedrich-Peinke Phenomenology to Phenomenological Models of Turbulence

In this section, we describe the relation of the Kramers-Moyal expansion (5.2.25) to scaling solutions of different phenomenologies of turbulence [26]. To this end, we take the moments of the one-increment PDF in Eq. (5.2.25)

$$
\frac{\partial}{\partial r}\langle v^n\rangle = \frac{\partial}{\partial r}\int_{-\infty}^{\infty} dv \, v^n f_1(v,r) = -\sum_{k=1}^{\infty}\int_{-\infty}^{\infty} dv \, v^n \left(-\frac{\partial}{\partial v}\right)^k D^{(k)}(v,r) f_1(v,r)
$$

$$
= -\sum_{k=1}^{n}\frac{n!}{(n-k)!}\int_{-\infty}^{\infty} dv \, v^{n-k} D^{(k)}(v,r) f_1(v,r) , \tag{5.2.28}
$$

where we dropped the indices of v_1 and r_1. In order to match powers of v, we choose $D^{(k)}(v,r) = \tilde{D}^{(k)}(r)v^k$ and obtain

$$
\frac{\partial}{\partial r}\ln\langle v^n\rangle = -\sum_{k=1}^{n}\frac{n!}{(n-k)!}\tilde{D}^{(k)}(r) . \tag{5.2.29}
$$

Moreover, scaling solutions $\langle v^n\rangle \sim r^{\zeta_n}$ can be ensured by imposing $\tilde{D}^{(k)}(r) = (-1)^k \frac{K_k}{k!}\frac{1}{r}$, which yields

$$
\frac{\partial}{\partial r}\ln\langle v^n\rangle = -\sum_{k=1}^{n}(-1)^k \binom{n}{k}\frac{1}{r}K_k . \tag{5.2.30}
$$

Integrating this equation from small scales r to the integral length scale L yields the longitudinal structure function of order n

$$
\langle v^n\rangle = \langle v_L^n\rangle r^{\sum_{k=1}^{n}(-1)^{k+1}\binom{n}{k}K_k} , \tag{5.2.31}
$$

where $\langle v_L^n\rangle = \int_{-\infty}^{\infty} dv \, v^n f_1(v,L)$ denotes the large-scale structure function. We can thus identify the exponent in Eq. (5.2.31) with the scaling exponent of the structure functions (see also Fig. 5.2)

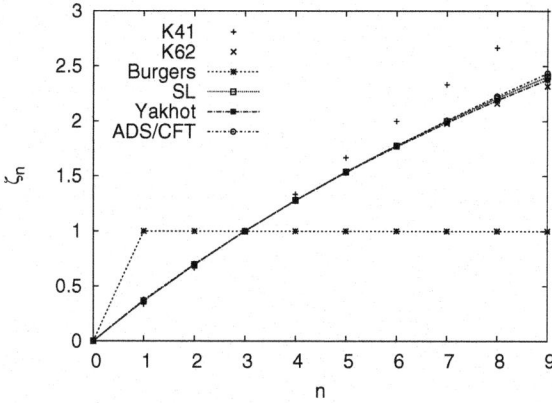

Fig. 5.2 Scaling exponents ζ_n of velocity structure functions for the different phenomenologies discussed in *(i.)–(vi.)*. The crosses that are arranged on the straight $n/3$-line correspond to the self-similar K41 phenomenology *(i.)*. Burgers phenomenology *(iii.)* exhibits the strongest intermittency behavior whereas other phenomenologies can only be distinguished for higher orders n. Note that the K62 phenomenology *(ii.)* has a parabolic form that violates the structure function convexity condition [6] for $n \geq \frac{3}{2} + \frac{3}{\mu}$ (not observable in the figure)

$$\zeta_n = \sum_{k=1}^{n} (-1)^{1-k} \binom{n}{k} K_k , \qquad (5.2.32)$$

where we will refer to K_k as the reduced Kramers-Moyal coefficients. Hence, Eq. (5.2.32) can be used as a recurrence relation and we can thus conclude that Kramers-Moyal coefficients of the form

$$D^{(n)}(v, r) = K_n \frac{(-1)^n}{n!} \frac{v^n}{r} \quad \text{and} \quad K_n = \sum_{k=1}^{n} (-1)^{1-k} \binom{n}{k} \zeta_k , \qquad (5.2.33)$$

necessarily imply scaling solutions $\langle v^n \rangle \sim r^{\zeta_n}$. The fact that the reduced Kramers-Moyal coefficients K_n are determined by the scaling exponents ζ_n shows that the Kramers-Moyal description is general enough to capture the essence of anomalous scaling. However, in its present form (5.2.33), the Kramers-Moyal expansion (5.2.25) is not capable of generating skewness during the cascade process from large to small scales. This can be readily seen from Eq. (5.2.31) which suggests that a non-vanishing third-order moment at scale r (in agreement with Kolmogorov's 4/5-law) requires the presence of a non-vanishing third-order moment at large scales $\langle v_L^3 \rangle$ and is thus at odds with empirical findings.

In the next subsections, we will describe how the different phenomenological models can be mapped onto Kramers-Moyal coefficients.

(i.) Kolmogorov's theory K41:
The self-similar K41 phenomenology [27] (see also Sect. 3.2.2) states that $\langle v^n \rangle = C_n \langle \varepsilon \rangle^{n/3} r^{n/3}$. An evaluation of the reduced Kramers-Moyal coefficients (5.2.33) suggests that it can be reproduced by just a single Kramers-Moyal coefficient

$$K_n = \begin{cases} 1/3 & \text{for } n \leq 1 \,, \\ 0 & \text{for } n > 1 \,. \end{cases} \tag{5.2.34}$$

(ii.) Kolmogorov-Oboukhov theory K62:
One of the first intermittency models which assumed a log-normal distribution of the local rate of energy dissipation ε was proposed by Kolmogorov [10] and Oboukhov [11] (see also Sect. 3.2.3.1). It predicts the scaling of the structure functions according to $\langle v^n \rangle = C_n \langle \varepsilon \rangle^{\frac{n}{3}} r^{\frac{n}{3}} \left(\frac{r}{L} \right)^{-\frac{n(n-3)\mu}{18}}$ where L is the integral length scale and μ is the so-called intermittency coefficient which is of the order $\mu \approx 0.227$. As discussed by Friedrich and Peinke [16], the K62 scaling reduces the Kramers-Moyal expansion to a Fokker-Planck equation with drift and diffusion coefficient

$$K_1 = \frac{3 + \mu}{9} \quad \text{and} \quad K_2 = \frac{\mu}{9} \,, \tag{5.2.35}$$

and implies the vanishing of all higher order coefficients.

(iii.) Burgers scaling:
The velocity structure functions in Burgers turbulence [28] follow the extreme scaling

$$\langle v^n \rangle = \begin{cases} C_n \frac{\langle \varepsilon^{n/2} \rangle}{\nu^{n/2}} r^n & \text{for } n < 1 \,, \\ C_n \langle \varepsilon \rangle^{\frac{n}{3}} L^{\frac{n}{3} - 1} r & \text{for } n \geq 1 \,. \end{cases} \tag{5.2.36}$$

Here, the first scaling is due to smooth positive velocity increments in the ramps, whereas the second scaling corresponds to negative velocity increments dominated by shocks that form due to the compressibility of the velocity field in the vicinity of the viscosity $\nu \to 0$. Smooth solutions correspond to a single Kramers-Moyal coefficient, whereas shock solutions can only be reproduced by an infinite number of Kramers-Moyal coefficients and we obtain

$$K_1 = 1, \quad K_n = 0 \text{ for } n > 1, \text{ for positive increments.}$$
$$K_n = 1, \text{ for negative increments .} \qquad (5.2.37)$$

(iv.) She-Leveque model:
The She-Leveque model [29] for 3D Navier-Stokes turbulence predicts scaling exponents $\zeta_n = \frac{n}{9} + 2\left(1 - \left(\frac{2}{3}\right)^{n/3}\right)$ that correspond well with both experimental and numerical data. This yields an infinite set of coefficients and the reduced Kramers-Moyal coefficients read [30]

$$K_n = \frac{n}{9}\,_1F_0(1 - n; ; 1) + 2\left(1 - \sqrt[3]{\frac{2}{3}}\right)^n , \qquad (5.2.38)$$

where $_\nu F_n(a; b; z)$ is the generalized hypergeometric function.

(v.) Yakhot model:
Yakhot [31–33] introduced a model for structure function exponents $\zeta_{2n} = \frac{2(1+3\beta)n}{3(1+2\beta n)}$ based on a mean-field approximation. Hence, Yakhot's model is to some degree derived directly from the Navier-Stokes equation with some additional assumptions, i.e., the validity of fusion rules and arguments invoked by the renormalization group [31]. Similar scaling exponents were first derived by Novikov [34] and subsequently by Castaing [35]. With the choice of $\beta = 0.05$, structure functions agree well with experimental data. The translation to the reduced Krames-Moyal coefficients is given by

$$K_n = \frac{\Gamma[n+1]}{\Gamma\left[n+1+\frac{1}{\beta}\right]}\left(\Gamma\left[1+\frac{1}{\beta}\right] + \frac{1}{3\beta^2}\Gamma\left[\frac{1}{\beta}\right]\right) . \qquad (5.2.39)$$

(vi.) ADS/CFT random geometry model:
Eling and Oz [36] introduced a structure function scaling model which is motivated by a gravitational Knizhnik-Polyakov-Zamolodchikov (KPZ)-type relation. For 3D Navier-Stokes turbulence, they derive

$$\zeta_n = \frac{\left((1+\gamma^2)^2 + 4\gamma^2(\frac{n}{3}-1)\right)^{\frac{1}{2}} + \gamma^2 - 1}{2\gamma^2} , \qquad (5.2.40)$$

Fig. 5.3 **a** Reduced
Kramers-Moyal coefficients
from Eq. (5.2.33) for
different phenomenological
models of turbulence up to
the order $n = 10$.
Coefficients for $n > 2$ seem
to tend toward zero. **b**
Semi-logarithmic plot of the
reduced Kramers-Moyal
coefficients. All
phenomenological models
except for K41 and K62
show an asymptotic
behavior. Note that the
She-Leveque possesses a
nearly linear slope in the
semi-logarithmic
representation

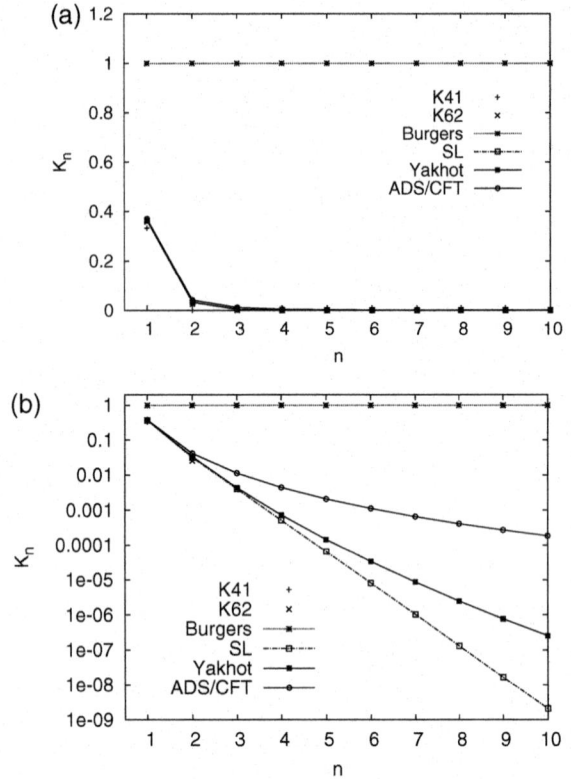

where experimental data suggests the value $\gamma^2 = 0.161$. Unfortunately, we could not
obtain an analytical formula for the coefficients of this particular model, and hence
narrowed our focus down to a numeric evaluation of Eq. (5.2.33).

The reduced Kramers-Moyal coefficients K_n for the different models up to the
order $n = 10$ are plotted in Fig. 5.3a. As one can see, all models except for K41 and
Burgers can be hardly distinguished from one another and the reduced Kramers-
Moyal coefficients seem to tend toward zero very quickly. According to a theorem
by Pawula [37] (see also [18]), the vanishing of the fourth-order Kramers-Moyal
coefficient implies that all higher coefficients are zero as well, and the Kramers-
Moyal expansion (5.2.25) reduces to an ordinary Fokker-Planck equation. The latter
is particularly suitable for modeling approaches due to its corresponding Langevin
equation as well as due to the undemanding determination of statistical quantities
via the exact short-scale propagator of the Fokker-Planck equation [18].

In the original work [16] and also in all subsequent works, it was argued in favor
of Pawula's theorem since the experimentally determined Kramers-Moyal coefficient of order four was very close to zero [19, 20, 22]. Figure 5.3a seems to agree
qualitatively with this finding. However, in order to demonstrate that this can be mis-
leading, Fig. 5.3b shows a semi-logarithmic plot of Fig. 5.3a: the models *(iv.)–(vi.)*

tend asymptotically toward zero and higher order Kramers-Moyal are rather small but strictly non-zero. At this point, we want to emphasize that since $K_4 \approx 10^{-3}$, the significant detection of these higher order coefficients in the experiment is quite challenging due to the presence of measurement noise or insufficient statistics. Nevertheless, since the models *(iv.)–(vi.)* correspond quite well with experimental data, an accurate determination of higher order coefficients should be within reach of a spatially and temporarily well-resolved high Reynolds number experiment. Moreover, Pawula's theorem directly reduces the velocity increment statistics to families of the K62 phenomenology *(ii.)*. It should therefore be noted that the latter is only valid for moments $\langle v^n \rangle$ that do not exceed the order $n \geq \frac{3}{2} + \frac{3}{\mu}$, because of the convexity condition for ζ_{2n} (see also [6] for further discussion). Consequently, one should bear in mind that in the course of modeling with the Fokker-Planck approach, the tails of PDFs might not be described accurately, although—admittedly—this effect should be rather small.

5.2.3 Fusion Rules and the Friedrich-Peinke Phenomenology

In the preceding section, we have investigated the implications of the Markov property (5.2.4) for the scaling of structure functions. In this section, we want to investigate the implications for multi-increment expectation values similar to the one discussed in Sect. 5.1 on fusion rules. Fusion rules suggest that the three-point correlation,

$$\langle v^j v'^k \rangle = \langle v(r, x)^j v(r', x)^k \rangle , \tag{5.2.41}$$

can be expressed in terms of the structure functions

$$S_n(r) = \langle v(r, x)^n \rangle , \tag{5.2.42}$$

according to

$$\langle v^j v'^k \rangle \sim \frac{S_j(r)}{S_j(r')} S_{j+k}(r') , \tag{5.2.43}$$

for $r < r'$ and $r' - r > \lambda_{fuse}$, where λ_{fuse} is a small-scale quantity that has yet to be determined.

On the other hand, the Markov property (5.2.4) suggests that the one-increment PDF follows the same Kramers-Moyal expansion as the transition PDF

$$-\frac{\partial}{\partial r} f_1(v, r) = \hat{L}_{KM}(v, r) f(v, r) , \tag{5.2.44}$$

$$-\frac{\partial}{\partial r} p(v, r | v', r') = \hat{L}_{KM}(v, r) p(v, r | v', r') . \tag{5.2.45}$$

where we defined the Kramers-Moyal operator

$$\hat{L}_{KM}(v, r) = \sum_{n=1}^{\infty} \left(-\frac{\partial}{\partial v} \right)^n D^{(n)}(v, r) ,\qquad (5.2.46)$$

with Kramers-Moyal coefficients

$$D^{(n)}(v', r') = \frac{1}{n!} \lim_{r \to r'} \frac{1}{r' - r} \int dv (v - v')^n p(v, r|v', r') .\qquad (5.2.47)$$

In order to impose scaling solutions for the structure functions, i.e., $S_n(r) \sim r^{\zeta_n}$, we focus on Kramers-Moyal coefficients of the form

$$D^{(n)}(v, r) = \frac{(-1)^n K_n}{n!} \frac{v^n}{r} .\qquad (5.2.48)$$

It can be shown from Eq. (5.2.33) that the reduced Kramers-Moyal coefficients K_n are related to the scaling exponents ζ_n according to

$$K_n = \sum_{k=1}^{n} (-1)^{1-k} \binom{n}{k} \zeta_k \quad \leftrightarrow \quad \zeta_n = \sum_{k=1}^{n} (-1)^{1-k} \binom{n}{k} K_k .\qquad (5.2.49)$$

The solution of Eq. (5.2.45) is expressed in form of a Dyson series [18]

$$p(v, r|v', r')$$
$$= \delta(v - v') + \int_r^{r'} dr_1 \hat{L}_{KM}(v, r_1) \delta(v - v')$$
$$+ \int_r^{r'} dr_1 \int_r^{r_1} dr_2 \hat{L}_{KM}(v, r_1) \hat{L}_{KM}(v, r_2) \delta(v - v') + \dots$$
$$= \delta(v - v') + \int_r^{r'} dr_1 \frac{\hat{L}}{r_1} \delta(v - v') + \int_r^{r'} dr_1 \int_r^{r_1} dr_2 \frac{\hat{L}^2}{r_1 r_2} \delta(v - v') + \dots$$
$$= \delta(v - v') + \ln \frac{r'}{r} \hat{L} \delta(v - v') + \frac{1}{2!} \left(\ln \frac{r'}{r} \right)^2 \hat{L}^2 \delta(v - v') + \dots$$
$$= \exp \left[\ln \frac{r'}{r} \hat{L} \right] \delta(v - v') ,\qquad (5.2.50)$$

where the differential operator $\hat{L}$ is defined according to

$$\hat{L} = \sum_{n=1}^{\infty} \frac{K_n}{n!} \frac{\partial^n}{\partial v^n} v^n \ . \tag{5.2.51}$$

We are now in the position to define three-point moments (5.2.41). Since $r < r'$, we can take the moments of the two-increment PDF $f_2(v, r; v', r') = p(v, r|v', r') f(v', r')$ and obtain

$$\langle v^j v'^k \rangle = \int dv\, v^j \int dv'\, v'^k \, p(v, r|v', r') f(v', r') \ . \tag{5.2.52}$$

Inserting the Dyson series (5.2.50) for the transition PDF p yields

$$\langle v^j v'^k \rangle$$

$$= \langle v'^{j+k} \rangle + \ln \frac{r'}{r} \sum_{n=1}^{\infty} \frac{K_n}{n!} \int dv' v'^k \int dv\, v^j \frac{\partial^n}{\partial v^n} v^n \delta(v - v') f_1(v', r')$$

$$+ \frac{1}{2!} \left(\ln \frac{r'}{r} \right)^2 \sum_{n,m=1}^{\infty} \frac{K_n K_m}{n! m!}$$

$$\times \int dv' v'^k \int dv\, v^j \frac{\partial^n}{\partial v^n} v^n \frac{\partial^m}{\partial v^m} v^m \delta(v - v') f_1(v', r') + \dots$$

$$= \langle v'^{j+k} \rangle + \ln \frac{r'}{r} \sum_{n=1}^{j} \frac{(-1)^n K_n j!}{n!(j-n)!} \int dv' v'^k \int dv\, v^j \delta(v - v') f_1(v', r')$$

$$+ \frac{1}{2!} \left(\ln \frac{r'}{r} \right)^2 \sum_{n=1}^{j} \frac{(-1)^n K_n j!}{n!(j-n)!} \sum_{m=1}^{j} \frac{(-1)^m K_m j!}{m!(j-m)!}$$

$$\times \int dv' v'^k \int dv\, v^j \delta(v - v') f_1(v', r') + \dots$$

$$= \langle v'^{j+k} \rangle + \ln \frac{r'}{r} \sum_{n=1}^{j} (-1)^n K_n \binom{j}{n} \langle v'^{j+k} \rangle$$

$$+ \frac{1}{2!} \left(\ln \frac{r'}{r} \right)^2 \sum_{n=1}^{j} (-1)^n K_n \binom{j}{n} \sum_{n'=1}^{j} (-1)^{n'} K_{n'} \binom{j}{n'} \langle v'^{j+k} \rangle + \dots$$

$$= \langle v'^{j+k} \rangle \left[1 - \ln \frac{r'}{r} \zeta_j + \frac{1}{2!} \left(\ln \frac{r'}{r} \right)^2 \zeta_j^2 + \dots \right]$$

$$= \langle v'^{j+k} \rangle \exp \left[\zeta_j \ln \frac{r}{r'} \right] = \langle v'^{j+k} \rangle \frac{r^{\zeta_j}}{r'^{\zeta_j}} = S_{j+k}(r') \frac{S_j(r)}{S_j(r')} \ . \tag{5.2.53}$$

Here, we performed partial integrations from the second to the third line. Subsequently, we applied relation (5.2.49) and inserted $S_j(r) \sim |r|^{\zeta_j}$ in the last step. This obviously is the same relation than the one suggested by the fusion rules (5.2.43).

Fusion rules are, therefore, a direct consequence of the Markov property (5.2.4). However, whether the inverse is true, i.e., if the fusion rules result in a Markov property of the velocity increments in scale, is far from obvious. Furthermore, the existence of a Markov-Einstein length λ_{ME} discussed in Sect. 5.2.1.2 implies that the fusion rules (5.2.43) are only valid in the inertial range and for $r' - r \leq \lambda_{fuse} = \lambda_{ME}$. We can thus summarize

Equivalence between fusion rules/operator product expansion and Markov processes of velocity increments in scale:

The fusion rules, which were discussed in Sect. 5.1, namely,

$$\langle v^j v'^k \rangle \sim \frac{\langle v^j \rangle}{\langle v'^j \rangle} \langle v'^{j+k} \rangle \,, \tag{5.2.54}$$

are a direct consequence of the Markov property (5.2.4) provided that structure functions exhibit scaling in the inertial range $\langle v^j \rangle \sim r^{\zeta_j}$. The latter two requirements are tantamount to Kramers-Moyal expansions (5.2.25–5.2.26) with Kramers-Moyal coefficients specified by Eq. (5.2.33).

5.2.4   Special Solutions for Transition Probabilities

In the following, we want to discuss solutions of the transition PDF for certain phenomenological models of turbulence. The point of departure is the Kramers-Moyal expansion for the transition PDF (5.2.26).

5.2.4.1   Solutions for the Transition Probabilities of Burgers Turbulence

The Burgers Kramers-Moyal expansion for the transition PDF of the shocks reads

$$\frac{\partial}{\partial r} p(v, r | v', r') = -\sum_{n=1}^{\infty} \frac{1}{n!} \frac{\partial^n}{\partial v^n} \frac{v^n}{r} p(v, r | v', r') \,. \tag{5.2.55}$$

A solution of this equation can be obtained from its Dyson series representation (5.2.50) according to

$$p(v, r | v', r') = \exp\left[\ln \frac{r}{r'} \hat{L}\right] \delta(v - v') \,, \tag{5.2.56}$$

where we introduced the operator

$$\hat{L} = -\sum_{n=1}^{\infty} \frac{1}{n!} \frac{\partial^n}{\partial v^n} v^n .$$ (5.2.57)

We can now let this operator act on the delta function and obtain

$$\hat{L}\delta(v - v') = -\sum_{n=1}^{\infty} \frac{1}{n!} \frac{\partial^n}{\partial v^n} v'^n \delta(v - v') ,$$ (5.2.58)

where we put the sifting property of the delta function into use. Now, we can write the delta function in its Fourier representation and obtain

$$\hat{L}\delta(v - v') = -\sum_{n=1}^{\infty} \frac{1}{n!} \frac{\partial^n}{\partial v^n} v'^n \int \frac{du}{2\pi} e^{iu(v-v')} = -\sum_{n=1}^{\infty} \int \frac{du}{2\pi} \frac{(iuv')^n}{n!} e^{iu(v-v')}$$

$$= \int \frac{du}{2\pi} e^{iu(v-v')} - \int \frac{du}{2\pi} e^{iuv'} e^{iu(v-v')} = \delta(v - v') - \delta(v) .$$ (5.2.59)

Another application of the operator yields

$$\hat{L}^2\delta(v - v') = \hat{L}\delta(v - v') - \hat{L}\delta(v) = -\delta(v - v') + \delta(v) .$$ (5.2.60)

Inserting this result into Eq. (5.2.56) yields

$$p(v, r|v', r') = \delta(v - v') - \ln\frac{r}{r'}(\delta(v - v') - \delta(v))$$ (5.2.61)

$$+ \frac{1}{2!}\left(\ln\frac{r}{r'}\right)^2 (\delta(v) - \delta(v - v')) + \ldots = e^{\ln\frac{r}{r'}}\delta(v - v') + (1 - e^{\ln\frac{r}{r'}})\delta(v) .$$

The transition PDF for negative increments v of the Burgers phenomenology thus reads

$$p(v, r|v', r') = \frac{r}{r'}\delta(v - v') + \left(1 - \frac{r}{r'}\right)\delta(v) .$$ (5.2.62)

It can be verified that this solution yields the correct Kramers-Moyal coefficients

$$D^{(n)}(v', r') = \frac{1}{n!} \lim_{r \to r'} \frac{1}{r' - r} \int dv(v - v')^n p(v, r|v', r')$$

$$= \frac{1}{n!} \int dv(v - v')^n \left(\frac{1}{r'}\delta(v - v') - \frac{1}{r'}\delta(v)\right) = -\frac{(-1)^n}{n!}\frac{v'^n}{r'} .$$ (5.2.63)

At the same time, it is also provable that the transition PDF (5.2.62) satisfies the coincidence property (5.2.9)

$$\lim_{r \to r'} p(v, r|v', r') = \delta(v - v') .$$

$$(5.2.64)$$

The Fokker-Planck equation for the transition PDF of positive velocity increments in Burgers turbulence reads

$$\frac{\partial}{\partial r} p(v, r|v', r') = -\frac{\partial}{\partial v} \frac{v}{r} p(v, r|v', r') ,$$

$$(5.2.65)$$

which is a first-order partial differential equation. The method of characteristics [38] suggests that we write the transition PDF in Eq. (5.2.65) in dependence of the parameter λ as $p(v(\lambda), r(\lambda)|v', r')$ which can be derived with respect to λ according to

$$\frac{d}{d\lambda} p(v(\lambda), r(\lambda)|v', r') = \frac{\partial p(v(\lambda), r(\lambda)|v', r')}{\partial r(\lambda)} \dot{r}(\lambda) + \frac{\partial p(v(\lambda), r(\lambda)|v', r')}{\partial v(\lambda)} \dot{v}(\lambda) .$$

$$(5.2.66)$$

Comparing this to Eq. (5.2.65) we obtain the following ordinary differential equations:

$$\dot{r}(\lambda) = 1 \quad \dot{v}(\lambda) = \frac{v(\lambda)}{r(\lambda)} \quad \frac{dp(v(\lambda), r(\lambda)|v', r')}{d\lambda} = -\frac{p(v(\lambda), r(\lambda)|v', r')}{r(\lambda)} .$$

$$(5.2.67)$$

Integrating the second and the third equation from r to r' with the initial condition $\delta(v - v')$ for p yields

$$v(\lambda) = v \frac{r(\lambda)}{r'} \quad p(v(\lambda), r(\lambda)|v', r') = \delta(v - v') \frac{r'}{r(\lambda)} .$$

$$(5.2.68)$$

Therefore, the transition for positive velocity increments v of the Burgers phenomenology reads

$$p(v, r|v', r') = \delta \left(v - \frac{r}{r'} v' \right) .$$

$$(5.2.69)$$

Again, the transition PDF yields the correct Kramers-Moyal coefficients (5.2.27) and satisfies the coincidence property (5.2.9).

5.2.4.2 Solution for the Transition Probability of the K41 Phenomenology

The Fokker-Planck equation for the transition PDF of the K41 phenomenology (5.2.34) reads

$$\frac{\partial}{\partial r} p(v, r | v', r') = -\frac{\partial}{\partial v} \frac{v}{3r} p(v, r | v', r') . \tag{5.2.70}$$

Again, this equation can be solved with the method of characteristics and we reach the system of equations

$$\dot{r}(\lambda) = 1 \quad \dot{v}(\lambda) = \frac{v(\lambda)}{3r(\lambda)} \quad \frac{\mathrm{d} p(v(\lambda), r(\lambda) | v', r')}{\mathrm{d}\lambda} = -\frac{p(v(\lambda), r(\lambda) | v', r')}{3r(\lambda)} , \tag{5.2.71}$$

which has the solution

$$p(v, r | v', r') = \delta \left(v - \frac{r^{1/3}}{r'^{1/3}} v' \right) . \tag{5.2.72}$$

It should be noted that this result is not in accordance with the large-scale limit $r' \gg r$, where the transition PDF should tend to a Gaussian distribution (the marginal distribution $f(v, r')$).

5.2.4.3 Solution for the Transition Probability of the K62 Phenomenology

The K62 phenomenology was already discussed in Sect. 5.2.2 under *(ii.)* and is equivalent to a Fokker-Planck equation

$$\frac{\partial}{\partial r} p(v, r | v', r') = \left[-\frac{\partial}{\partial v} \frac{3 + \mu}{9r} v - \frac{\partial^2}{\partial v^2} \frac{\mu}{18r} v^2 \right] p(v, r | v', r') . \tag{5.2.73}$$

We introduce

$$A = \frac{3 + \mu}{9} \quad B = -\frac{\mu}{18} , \tag{5.2.74}$$

and choose our ansatz as a log-normal distribution of the form

$$p(v, r | v', r') = \frac{1}{\sqrt{2\pi Q(r, r')} v} \exp \left[-\frac{\left(\ln \frac{v}{v'} - K(r, r') \right)^2}{? Q(r, r')} \right] , \tag{5.2.75}$$

where $K(r, r')$ and $Q(r, r')$ are functions that are yet to be determined by Eq. (5.2.73). Deriving Eq. (5.2.75) with respect to r yields

$$\frac{\partial}{\partial r} p(v, r|v', r') = \left[-\frac{\dot{Q}}{2Q} + \frac{\left(\ln \frac{v}{v'} - K \right)}{Q} \dot{K} + \frac{\left(\ln \frac{v}{v'} - K \right)^2}{2Q^2} \dot{Q} \right] p(v, r|v', r') ,$$

(5.2.76)

where the dot indicates a derivative with respect to r.
The r.h.s. of Eq. (5.2.73) is evaluated as follows:

$$\frac{\partial}{\partial v} v p(v, r|v', r') = -\frac{\left(\ln \frac{v}{v'} - K \right)}{Q} p(v, r|v', r') ,$$

(5.2.77)

and

$$\frac{\partial^2}{\partial v^2} v^2 p(v, r|v', r') = \left[-\frac{1}{Q} - \frac{\left(\ln \frac{v}{v'} - K \right)}{Q} + \frac{\left(\ln \frac{v}{v'} - K \right)^2}{Q^2} \right] p(v, r|v', r') .$$

(5.2.78)

The determining equation for $K(r, r')$ and $Q(r, r')$ thus reads

$$-\frac{\dot{Q}}{2Q} + \frac{\left(\ln \frac{v}{v'} - K \right)}{Q} \dot{K} + \frac{\left(\ln \frac{v}{v'} - K \right)^2}{2Q^2} \dot{Q}$$

(5.2.79)

$$= -\frac{B}{Qr} + \left(\frac{A}{r} - \frac{B}{r} \right) \frac{\left(\ln \frac{v}{v'} - K \right)}{Q} + \frac{B}{r} \frac{\left(\ln \frac{v}{v'} - K \right)^2}{Q^2} ,$$

(5.2.80)

which implies the following ordinary differential equations for $K(r, r')$ and $Q(r, r')$:

$$\dot{Q}(r, r') = \frac{2B}{r} , \quad \text{and} \quad \dot{K}(r, r') = \left(\frac{A}{r} - \frac{B}{r} \right) .$$

(5.2.81)

These ODEs can be integrated according to

$$Q(r, r') = 2b \ln \frac{r}{r'} , \quad \text{and} \quad K(r, r') = a \ln \frac{r}{r'} ,$$

(5.2.82)

where

$$a = A - B = \frac{1}{3} + \frac{\mu}{6} , \quad \text{and} \quad b = -B = \frac{\mu}{18} .$$

(5.2.83)

The exact solution for the transition probability of this particular Fokker-Planck equation of the K62 phenomenology then reads

$$p(v, r|v', r') = \frac{1}{\sqrt{4\pi b \ln \frac{r}{r'}} v} \exp \left[-\frac{\left(\ln \frac{v}{v'} - a \ln \frac{r}{r'} \right)^2}{4b \ln \frac{r}{r'}} \right] .$$

(5.2.84)

We find that for $r \to r'$, the transition probability approaches a delta function, according to

$$\lim_{r \to r'} p(v, r | v', r') = \frac{1}{v} \delta \left(\ln \frac{v}{v'} \right) = \delta(v - v') , \qquad (5.2.85)$$

which is in accordance with the coincidence property (5.2.9).

In the following, we investigate the relation of the transition PDF (5.2.84) to the one-increment PDF proposed by Yakhot [33] as well as by Castaing [39]. In using the so-called Mellin transform, Yakhot was able to derive the one-increment PDF directly from the structure functions of the K62 phenomenology (3.2.24). Furthermore, he assumed that the PDF follows a Gaussian distribution at large scales, e.g., for $r = 1$ he stated that $f_1(v, r = 1) = \frac{e^{-v^2/2}}{\sqrt{2\pi}}$. Yakhot's formula can be obtained from the transition PDF (5.2.84) in setting $r' = 1$ and making use of the reduction property of the two-increment PDF (5.2.8).

$$f_1(v, r) = \int dv' \, p(v, r | v', r' = 1) f_1(v', r' = 1)$$

$$= \frac{1}{\pi v \sqrt{8b \ln r}} \int dv' e^{-v'^2/2} \exp \left[-\frac{\left(\ln v - a \ln r - \ln v' \right)^2}{4b \ln r} \right] , \qquad (5.2.86)$$

which reduces to a Gaussian for $r = 1$, as demanded.

In analogy to the K62 phenomenology, Castaing's model of multiplicative energy cascades is based on local fluctuations of the energy dissipation rate, which follow a log-normal distribution. However, in contrast to K62 which predicts scaling of the structure functions, Castaing's formula,

$$f_1(v, r) = \frac{1}{2\pi \lambda(r)} \int \frac{ds}{s^2} \exp \left[-\frac{\ln^2(s/s_0(r))}{2\lambda^2(r)} \right] e^{-v^2/2s^2} , \qquad (5.2.87)$$

was devised to fit experimental data via the fitting functions $s_0(r)$ and $\lambda(r)$ and does not necessarily imply structure function scaling. Castaing's formula (5.2.87) is more general than Yakhot's formula (5.2.86), which can be recovered in substituting $s = v/v'$ and choosing $s_0(r) = r^a$ and $\lambda(r) = \sqrt{2b \ln r}$.

5.2.5   *Markov Analysis of Turbulence Data

In this section, we want to outline a systematic Markov analysis of turbulence data at the example of direct numerical simulations of three-dimensional turbulence [40, 41]. The methods presented here, however, can also be applied to any turbulent signal provided that it contains a sufficient number of data points. The analysis can be summarized as follows:

Principal steps of the Markov analysis of turbulence data:

(a) Determine the one- and two-time conditional PDF of longitudinal velocity increments (5.2.1) using Bayes' theorem (5.2.3).
(b) Examine the Markov property either directly (5.2.4) or by use of the Chapman-Kolmogorov equation (5.2.7).
(c) Quantify the similarities between the two PDFs (5.2.4) or (5.2.7) in more detail; introduce a distance measure and estimate a small-scale separation $\lambda_{ME} = r_2 - r_3$ for which the Markov property is violated.
(d) In order to estimate the Kramers-Moyal coefficients (5.2.27), calculate the conditional moments of the transition PDF $p(v_3, r_2 - \lambda | v_2, r_2)$ as a function of λ (for each v_2, r_2) and extrapolate them for $\lambda < \lambda_{ME}$.
(e) Project on the *universal features* of the turbulence data, i.e., the terms of the Kramers-Moyal coefficients (5.2.33) which entail scaling solutions (5.2.31).
(f) Investigate the asymptotics of higher order Kramers-Moyal coefficients in a plot similar to Fig. 5.3b.

As it will be discussed in the following sections, the proposed analysis can in some sense be considered as an alternative to the usual scaling exponent approach. The advantage of the Markov analysis is its ability to extract the scaling features of the turbulent signal. Moreover, as suggested by Fig. 5.3b, the reduced Kramers-Moyal coefficients K_n tend to zero or saturate, whereas the structure functions follow the convexity condition discussed in Sect. 3.2.3.1.

5.2.5.1 *Examination of the Markov Property

In the following, we analyze data obtained from a pseudospectral code developed for direct numerical simulations of the three-dimensional Navier-Stokes equation. For further details, we refer the reader to the PhD thesis of H. Homann [41] as well as to the paper which investigates longitudinal and transverse structure functions [42]. Here, we focus on data with a spatial resolution of 2048^3 grid points and a Taylor-Reynolds number $R_\lambda = 460$. The remaining characteristics of the simulations are summarized in Table 5.1. The Markov analysis starts with a thorough investigation and validation of the Markov property (5.2.4) of longitudinal velocity increments,

$$v(\mathbf{r}, \mathbf{x}, t) = (\mathbf{u}(\mathbf{x} + \mathbf{r}, t) - \mathbf{u}(\mathbf{x}, t)) \cdot \frac{\mathbf{r}}{r}, \qquad (*5.2.88)$$

which are easily accessible in three-dimensional turbulence (in the case of a single-probe turbulent measurement, e.g., using hot wire anemometry, these quantities can be obtained by the use of Taylor's hypothesis from Sect. 3.2.1.1). The evolution of

Table 5.1 Characteristic parameters of the direct numerical simulations of 3D hydrodynamic turbulence: Taylor-Reynolds number $Re_\lambda = \sqrt{\frac{15u_{rms}L}{\nu}}$, root mean square velocity $u_{rms} = \sqrt{\langle \mathbf{u}^2 \rangle}$, averaged kinetic energy dissipation rate $\langle \varepsilon \rangle$, kinematic viscosity ν, dissipation length $\eta = \left(\frac{\nu^3}{\langle \varepsilon \rangle}\right)^{1/4}$, dissipation time $\tau_\eta = \left(\frac{\nu}{\eta}\right)^{1/2}$, integral length scale L, large eddy turnover time $T_L = \frac{L}{u_{rms}}$, and resolution N. The length of the box is $L_{box} = 2\pi$

Re_λ	u_{rms}	$\langle \varepsilon \rangle$	ν	dx	η	τ_η	λ	L	T_L	N^3
460	0.189	$3.6 \cdot 10^{-3}$	$2.5 \cdot 10^{-5}$	$3.07 \cdot 10^{-3}$	$1.45 \cdot 10^{-3}$	0.083	0.06132	1.85	9.9	2048^3

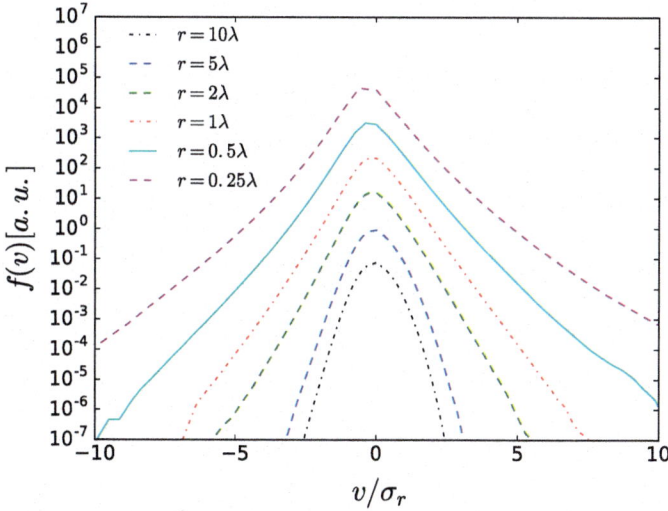

Fig. 5.4 Evolution of the velocity increment PDF in scale for DNS of three-dimensional turbulence. The PDFs exhibit strong non-self-similar behavior in the inertial range

the longitudinal increment PDF $f_1(v, r)$ in scale is depicted in Fig. 5.4. The PDF exhibits pronounced tails for small-scale velocity increments, a typical feature of intermittency. In the following, we briefly mention two possible methods of validation of the Markov property. First of all, the Markov property can be examined directly via comparing the conditioned PDF to the transition PDF

$$p(v_3, L/2 - \Delta r | v_2, L/2; v_1, L/2 + \Delta r) = p(v_3, L/2 - \Delta r | v_2, L/2) , \quad (*5.2.89)$$

where the intermediate scale $L/2$ was chosen to lie well within the inertial range and Δr can be considered as the variable step width of the process. In general, the intermediate scale can also be chosen at a different scale in the inertial range; however, this particular case should ensure that both v_3 and v_1 lie within the inertial range. Furthermore, it is also possible to examine the generalized Markov property

(5.2.5). However, such testing requires considerably more significant points in the dataset due to the higher dimensionality of the conditional PDF on the l.h.s. of Eq. (5.2.5) in comparison to the two-time conditional PDF on the l.h.s. of Eq. (*5.2.89). Hence, in order to check the validity of Eq. (*5.2.89) one first needs to evaluate the two- and three-increment histograms, see Eq. (5.2.2). The conditional PDFs that enter Eq. (*5.2.89) are then determined according to

$$p(v_3, L/2 - \Delta r | v_2, L/2; v_1, L/2 + \Delta r) = \frac{f_3(v_3, L/2 - \Delta r; v_2, L/2; v_1, L/2 + \Delta r)}{f_2(v_2, L/2; v_1, L/2 + \Delta r)}$$

$$p(v_3, L/2 - \Delta r | v_2, L/2) = \frac{f_2(v_3, L/2 - \Delta r; v_2, L/2)}{f_1(v_2, L/2)} . \quad (*5.2.90)$$

Latter procedure involves a technicality which is due to increment values v_3, v_2, and v_1, for which the denominators in Eqs. (*5.2.90) are zero. The most convenient way to treat these cases is simply to set the corresponding conditional probabilities to zero. Another difficulty is that Eq. (*5.2.89) compares two objects of different dimensionality. Consequently, we have to perform cuts in the two-time conditioned PDF for fixed v_1. This can be done by the simple choice $v_1 = 0$, which is the most significant case. In order to test the Markov property for other values of v_1, it is convenient to introduce a scale-independent quantity, i.e., the standard deviation of the velocity increment (*5.2.88) at large scales [20]

$$\sigma_\infty = \lim_{r \to \infty} \sqrt{\langle v(\mathbf{x}, r)^2 \rangle} = \lim_{r \to \infty} \sqrt{\left\langle \left[(u(\mathbf{x} + \mathbf{r}) - u(\mathbf{x})) \cdot \frac{\mathbf{r}}{r} \right]^2 \right\rangle} = \sqrt{2} u_{rms} . \quad (*5.2.91)$$

Therefore, it may be appropriate to test the Markov property (*5.2.89) for different significance levels. Figure 5.5 shows the examination of the Markov property for the scale separation $\Delta r = 0.5\lambda$ for both $v_1 = 0$ and $v_1 = 1\sigma_\infty$. The Markov property is not fulfilled for large values of v_2, where the two-time conditioned PDFs (red) become more curved than the transition PDFs (blue). However, the Markov property seems to be rather accurate in the center of the PDFs. Increasing the scale separation leads to an improvement of the Markov property, which can be seen from Fig. 5.6 for the scale separation $\Delta r = \lambda$. Further increasing the scale separation improves the Markov property, for instance, in Fig. 5.7 for $\Delta r = 4.5\lambda$.

A different method to validate the Markov property is to check the Chapman-Kolmogorov equation (5.2.7)

$$\int dv_2\, p(v_3, L/2 - \Delta r | v_2, L/2)\, p(v_2, L/2 | v_1, L/2 + \Delta r)$$
$$= p(v_3, L/2 - \Delta r | v_1, L/2 + \Delta r) , \quad (*5.2.92)$$

which solely involves transition PDFs. Hence, validation is reduced to the comparison of two quantities which possess the same dimensionality (unlike the direct Markov test (*5.2.89)).

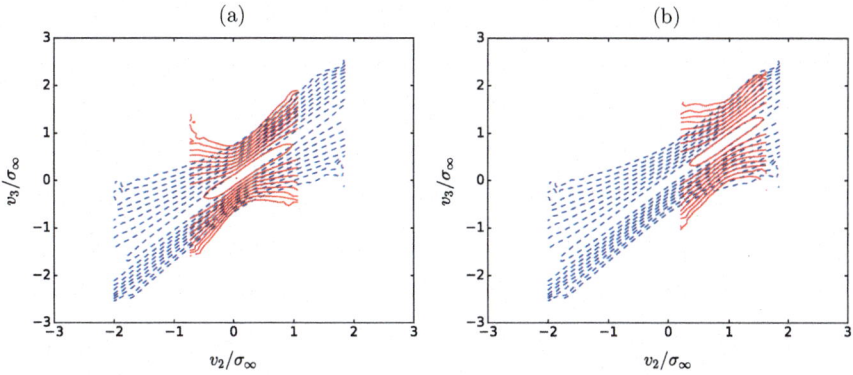

Fig. 5.5 a Examination of the Markov property (*5.2.89) from DNS of three-dimensional turbulence for $\Delta r = 0.5\lambda$ and $v_1 = 0$ via a logarithmic contour plot. The dashed blue contour lines correspond to $p(v_3, L/2 - \Delta r|v_2, L/2)$ whereas the red lines correspond to $p(v_3, L/2 - \Delta r|v_2, L/2; v_1, L/2 + \Delta r)$. The Markov property is violated in boundary regions. **b** Same as in (**a**), but for $v_1 = 1\sigma_\infty$

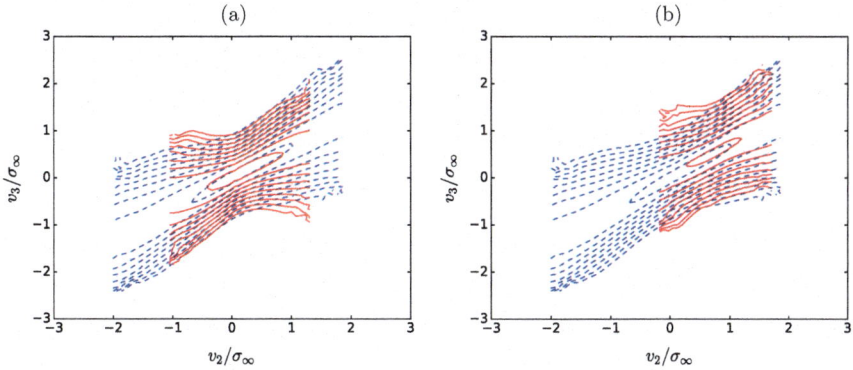

Fig. 5.6 *a Examination of the Markov property (*5.2.89) from DNS of three-dimensional turbulence for $\Delta r = \lambda$ and $v_1 = 0$ via a logarithmic contour plot. The dashed blue contour lines correspond to $p(v_3, L/2 - \Delta r|v_2, L/2)$ whereas the red lines correspond to $p(v_3, L/2 - \Delta r|v_2, L/2; v_1, L/2 + \Delta r)$. The Markov property is violated in the boundary regions, but still seems a good approximation in the center of the PDF. **b** Same as in (**a**), but for $v_1 = 1\sigma_\infty$

5.2.5.2 *Determination of the Markov-Einstein Length

Here, we seek to quantify the breakdown of the Markov property in more detail. To this end, we need some kind of distance measure between the one- and two-time conditional PDF in Eq. (*5.2.89). Several measures have been proposed, e.g., the chi-squared test [43], the Wilcoxon test [19, 20], the Kullback-Leibler entropy [20], or the Hellinger distance [44]. The advantage of the Hellinger distance H is that it forms a true metric in the space of the PDFs and be used to decide at which scale

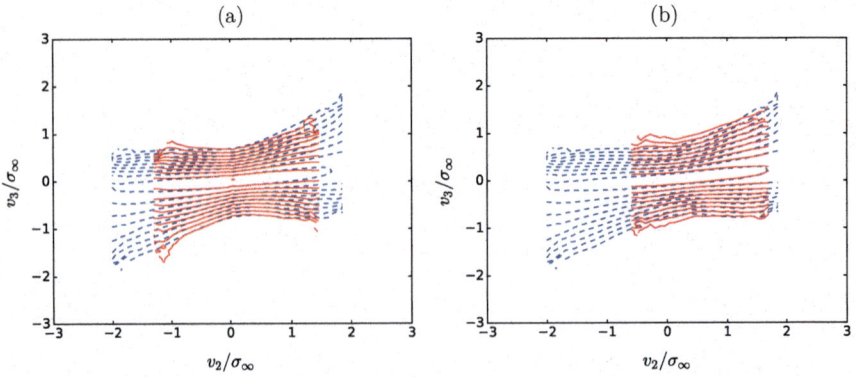

Fig. 5.7 a Examination of the Markov property (*5.2.89) from DNS of three-dimensional turbulence for $\Delta r = 4.5\lambda$ and $v_1 = 0$ via a logarithmic contour plot. The dashed blue contour lines correspond to $p(v_3, L/2 - \Delta r | v_2, L/2)$ whereas the red lines correspond to $p(v_3, L/2 - \Delta r | v_2, L/2; v_1, L/2 + \Delta r)$. The Markov property is fulfilled. **b** Same as in (**a**), but for $v_1 = 1\sigma_\infty$

separation Δr the Markov property significantly deteriorates. The Hellinger distance H for continuous distributions is defined according to

$$
H^2 \ (v_2, v_1; \Delta r) = \frac{1}{2} \int dv_3 \left(\sqrt{p(v_3, L/2 - \Delta r | v_2, L/2; v_1, L/2 + \Delta r)} \right.
$$
$$
\left. - \sqrt{p(v_3, L/2 - \Delta r | v_2, L/2)} \right)^2
$$
$$
= 1 - \int dv_3 \sqrt{p(v_3, L/2 - \Delta r | v_2, L/2; v_1, L/2 + \Delta r) p(v_3, L/2 - \Delta r | v_2, L/2)} \ .
$$
$$
(*5.2.93)
$$

Here, we made use of the identities

$$
\int dv_3 p(v_3, L/2 - \Delta r | v_2, L/2; v_1, L/2 + \Delta r) = 1 \qquad (*5.2.94)
$$

$$
\int dv_3 p(v_3, L/2 - \Delta r | v_2, L/2) = 1 \ , \qquad (*5.2.95)
$$

in the last step. Hence, the Hellinger distance is symmetric in both probabilities and is restricted to

$$
0 \le H(v_2, v_1; \Delta r) \le 1 \ , \qquad (*5.2.96)
$$

which is a direct consequence of the Cauchy-Schwarz inequality. Another useful property of the Hellinger distance is that it can be explicitly calculated for certain types of PDFs (normal distribution, beta distribution, exponential distribution, etc.).

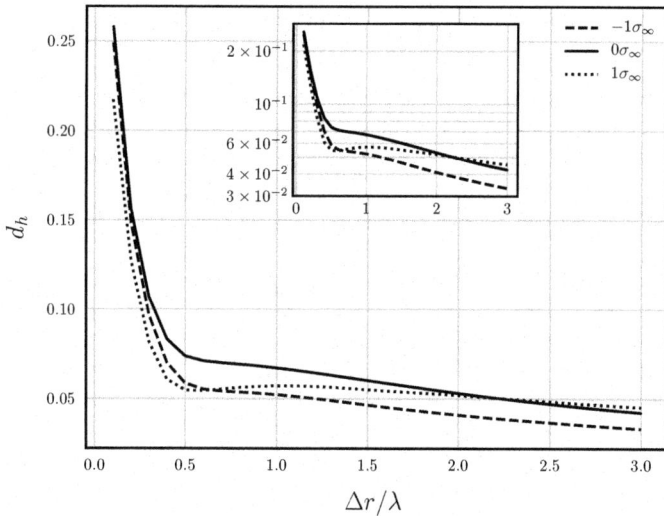

Fig. 5.8 Hellinger distance $d_H(\Delta r)$ for three different $v_1 = -1\sigma_\infty$, 0, and $1\sigma_\infty$, and variable step width Δr in unities of the Taylor length λ. The Hellinger distance increases considerably at around $\Delta r \approx 0.5\lambda$, whereas it exhibits a rather linear increase for $\Delta r > 0.5\lambda$. *Inset:* Semi-logarithmic plot of the Hellinger distance $d_H(\Delta r)$. For $\Delta r < 0.5\lambda$, the Hellinger distance seems to increase nearly exponentially

In our case, the Hellinger distance is still a function of v_2, if we assume that v_1 is fixed. Therefore, an average of the corresponding v_2-values is performed in order to obtain a pure correlation measure

$$d_H(\Delta r, v_1) = \langle H(v_2, v_1; \Delta r)\rangle_{v_2} . \qquad (*5.2.97)$$

The corresponding Δr-dependence of the Hellinger distance in Fig. 5.8a is expressed in terms of the Taylor length λ. It exhibits a pronounced increase for smaller values than $\Delta r \approx 0.5\lambda$. Hence, the Markov property is clearly violated at scale separations $\Delta r < 0.5\lambda$. Moreover, even for larger values of Δr, the Hellinger distance is still dropping to zero, which indicates that the Markov property is not fully fulfilled at these scale separations either. However, the close to exponential increase at $\Delta r \approx 0.5\lambda$ is more significant than the linear increase for higher values of Δr. We choose the Markov-Einstein length at $\lambda_{ME} = \lambda$. Nevertheless, it must be stressed that the Markov property is only fully fulfilled for $\Delta r > 3\lambda$. Hence, the task of an accurate determination of the Markov-Einstein length is far from obvious. However, it can be inferred from Fig. 5.8 that it lies near the Taylor length as the correlation measure clearly drops at $\Delta r/\lambda \approx 0.5$.

5.2.6   *Determination of the Kramers-Moyal Coefficients

In this section, we will outline a procedure which determines Kramers-Moyal coefficients (5.2.27) from numerically evaluated transition probabilities similar to Figs. 5.5, 5.6, 5.7. As the Markov property becomes violated in the proximity of the Einstein-Markov length, it is necessary to use an appropriate extrapolation for the conditional moments

$$M^{(n)}(v, r; \Delta r) = \frac{1}{n!} \int dv'(v' - v)^n p(v', r - \Delta r | v, r) . \qquad (*5.2.98)$$

Consequently, we expand the conditional moments (*5.2.98) in a Taylor series for small Δr

$$M^{(n)}(v, r; \Delta r) = D^{(n)}(v, r)\Delta r + \mathcal{O}(\Delta r^2) , \qquad (*5.2.99)$$

which is commonly referred to as Itô-Taylor expansion [45]. Subsequently, the limit,

$$\lim_{\Delta r \to 0} \frac{M^{(n)}(v, r; \Delta r)}{\Delta r} = D^{(n)}(v, r) , \qquad (*5.2.100)$$

has to be determined from extrapolating the conditional moments in order to obtain the corresponding Kramers-Moyal coefficient [20]. As an example, we plotted conditional moments of first order for four different v in Fig. 5.9. It can be seen that

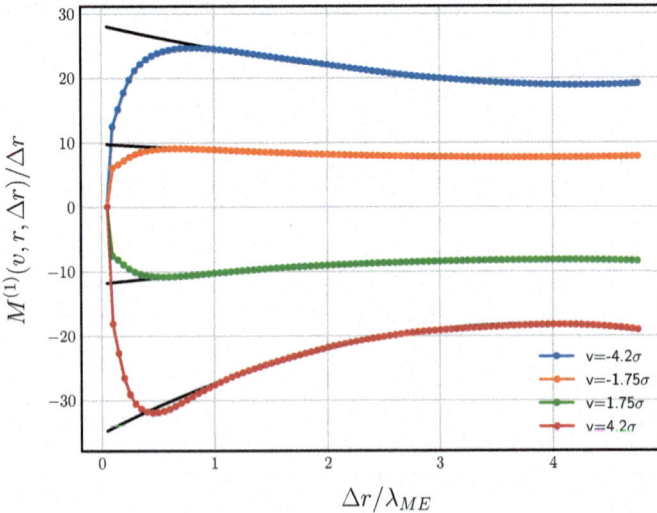

Fig. 5.9 Conditional moments of first order divided by scale separation Δr, $M^{(1)}(v, r; \Delta r)/\Delta r$ for $r = L/2$, and variable Δr. The fits (black lines) correspond to polynomials of second order in Δr for $\Delta r > \lambda_{ME}$. Note that the Δr-axis has been rescaled by the Markov-Einstein length λ_{ME} and the conditional moment drops to zero for $\Delta r < \lambda_{ME}$

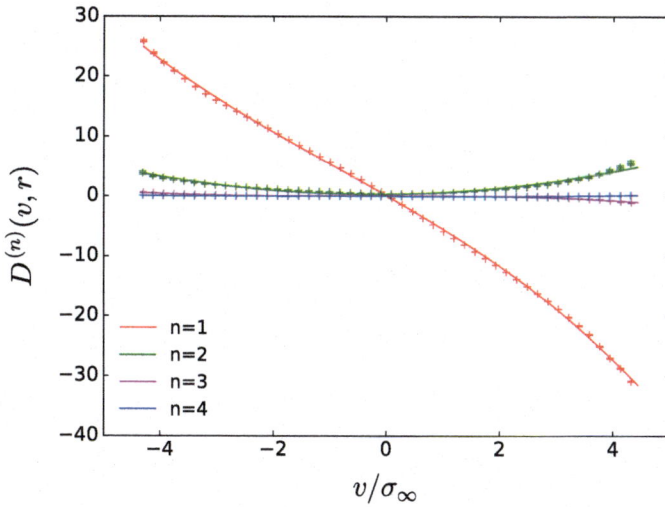

Fig. 5.10 Estimation of the Kramers-Moyal coefficients $D^{(n)}(v, r)$ from DNS of three-dimensional turbulence for $r = 5\lambda$. The fits correspond to polynomials of the order n of the coefficient except for $n = 1$ where a polynomial of order three has been used. The reduced Kramers-Moyal coefficients have been determined according to $K_1 = 0.265 \pm 0.005$, $K_2 = 0.02112 \pm 0.0007$, $K_3 = 0.0027 \pm 0.0001$, and $K_4 = 0.00052 \pm 0.00002$

$M^{(1)}(v, r; \Delta r)/\Delta r$ drops against zero for $\Delta r \lesssim \lambda_{ME}$ as the Markov property is violated. In order to extrapolate the moments, polynomial fits of second order in Δr were performed for $\Delta r > \lambda_{ME}$ (black lines in the plot). The Kramers-Moyal coefficient for a particular v, in this case the drift coefficient, can be determined from the y-intercepts of the fits. Consequently, this procedure has to be repeated for several v (and, in general, also different r) for the sake of obtaining the full functional form of the Kramers-Moyal coefficients.

Figure 5.10 shows the resulting Kramers-Moyal coefficients $D^{(n)}(v, r)$ for $r = 5\lambda$ up to order $n = 4$. All coefficients can be approximated by polynomials of order n, except for the drift coefficient which possesses an additional cubic dependence on v. Higher order coefficients are rather small but exhibit polynomial behavior. The reduced Kramers-Moyal coefficients (see Eq. (5.2.48)) evaluated at this particular scale $r = 5\lambda$ are depicted in Fig. 5.11. Here, averages over one, three, and five time snapshots of the velocity field have been performed. For higher n, the coefficients agree fairly well with Yakhot's phenomenology. Averaging over five time slices yields an almost perfect agreement with Yakhot's phenomenology for $n \geq 5$. Nevertheless, for $n < 5$, the reduced Kramers-Moyal coefficients from DNS slightly underestimate the coefficients predicted by Yakhot's model. In order to understand the cause for this peculiarity, we have to evaluate the Kramers-Moyal coefficients for variable scales r. Therefore, as far as the Markov analysis is concerned, it is appropriate to decompose the Kramers-Moyal coefficients according to

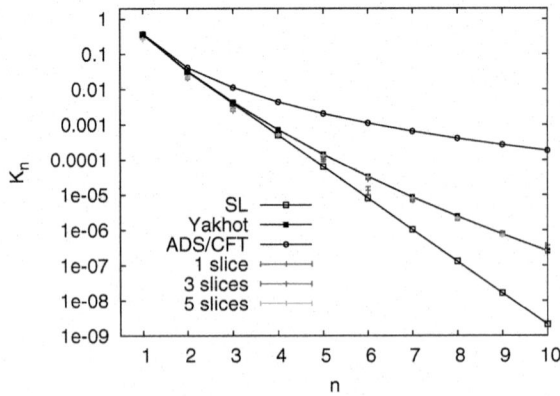

Fig. 5.11 Semi-logarithmic plot of the reduced Kramers-Moyal coefficients for a fixed $r = 5\lambda$ averaged over one, three, and five slices of the velocity field. Higher order coefficients agree remarkably well with Yakhot's model. Averaging over more slices further increases this tendency. Note that the average over a single slice (red) was not sufficient for the determination of reduced coefficients $n \geq 7$

$$D^{(n)}(v, r) = \underbrace{\frac{(-1)^n}{n!} K_n \frac{v^n}{r}}_{\text{scaling/universal part}} + \underbrace{\text{non-universal parts}}_{\text{boundary/finite Reynolds effects}} \quad . \qquad (*5.2.101)$$

Therefore, by projecting on the first term in Eq. (*5.2.101), in this case by simple polynomial fits in Fig. 5.11, one can extract valuable information on scaling features of turbulence.

5.3   The Instanton Method in Turbulence

Another interesting method that emanated from field theory is the so-called instanton appproach [46]. This method starts from a path integral formulation for stochastic partial differential equations which is commonly referred to as Martin-Siggia-Rose formalism [47] (see also [48]). Instantons are defined as the saddle point configuration of the corresponding path integrals.

5.3.1   The Martin-Siggia-Rose Formalism

We consider a Langevin-type equation of the form

$$\frac{\partial}{\partial t} u(x, t) + \mathcal{F}[u] = \eta(x, t) , \qquad (5.3.1)$$

where $\mathcal{F}$ is a functional operator that entirely characterizes the dynamics of the system. Moreover, $\eta(x, t)$ is a stochastic force that is white noise in time, $\langle \eta(x, t)\eta(x', t')\rangle = \chi(x - x')\delta(t - t')$, and possesses the spatial correlation function $\chi(x)$.

We can cast the expectation value of an observable $\mathcal{O}[u]$ as a path integral over all realizations of the noise

$$\langle \mathcal{O}[u]\rangle = \int \mathcal{D}\eta \, \mathcal{O}[u]e^{-\langle \eta, \chi^{-1}\eta\rangle/2} . \tag{5.3.2}$$

Here, the brackets in the exponent of the weighting factor indicate a suitable inner product, e.g., the L^2-norm. Via introducing the auxiliary field $\mu(x, t)$, we are able to express the weighting factor in terms of the original correlation function $\chi(x)$

$$\langle \mathcal{O}[u]\rangle = \int \mathcal{D}\mu \, \mathcal{D}\eta \, \mathcal{O}[u]e^{-\langle \mu, \chi\mu\rangle/2 + i\langle \mu, \eta\rangle} . \tag{5.3.3}$$

In the following, we want to pass from the noise integration to an integration over all realizations of the field $u(x, t)$, which results in an Onsager-Machlup functional [18, 49]. The field $u(x, t)$ depends implicitly on the forcing $\eta(x, t)$ due to Eq. (5.3.1). Therefore, the coordinate transform leads to a Jacobian in $\mathcal{D}\eta = J[u]\mathcal{D}u$, which is determined by

$$J[u] = \det\left(\frac{\delta\eta}{\delta u}\right) = \det\left(\frac{\partial}{\partial t} - \frac{\delta\mathcal{F}}{\delta u}\right) . \tag{5.3.4}$$

These manipulations lead to

$$\langle \mathcal{O}[u]\rangle = \int \mathcal{D}\mu\mathcal{D}u\mathcal{O}[u]J[u]e^{-S[u,\mu]} , \tag{5.3.5}$$

with the action functional

$$S[u, \mu] = -i\left\langle \mu, \frac{\partial}{\partial t}u - \mathcal{F}[u]\right\rangle + \frac{1}{2}\langle \mu, \chi\mu\rangle . \tag{5.3.6}$$

The instanton method was devised in order to calculate dominant contributions to the path integral (5.3.5) via a saddle point approximation, i.e., by finding the extremum of the action functional.

5.3.2 Extracting the Instanton

To be more specific, we choose the action functional of the stochastic Burgers equation, i.e., Eq. (2.4.1) with forcing

$$S[u, \mu] = -i \int dt \int dx \mu(x, t) \left(\frac{\partial}{\partial t} u(x, t) + u(x, t) \frac{\partial}{\partial x} u(x, t) - \nu \frac{\partial^2}{\partial x^2} u(x, t) \right)$$
$$+ \frac{1}{2} \int dt \int dx \int dx' \mu(x, t) \chi(x - x') \mu(x', t) . \qquad (5.3.7)$$

Hence, the velocity gradient PDF at time $t = 0$, for instance, can be expressed according to

$$g(a) = \left\langle \delta \left(a - \delta(t) \frac{\partial u(x, t)}{\partial x} \right) \right\rangle$$
$$= \int \mathcal{D}\mu \mathcal{D}u \, J[u] \int d\lambda \exp \left[-S[u, \mu] + i\lambda \left(a - \delta(t) \frac{\partial u(x, t)}{\partial x} \right) \right] . \qquad (5.3.8)$$

For a saddle point approximation of this integral, we have to determine functional derivatives of the effective action

$$S_{eff} = S + i\lambda \left(a - \delta(t) \frac{\partial u(x, t)}{\partial x} \right) , \qquad (5.3.9)$$

according to

$$\frac{\delta S_{eff}}{\delta u(x, t)} = 0 \quad \text{and} \quad \frac{\delta S_{eff}}{\delta \mu(x, t)} = 0 . \qquad (5.3.10)$$

Hence, minimization of the action results in the *instanton equations*.

$$\frac{\partial}{\partial t} u(x, t) + u(x, t) \frac{\partial}{\partial x} u(x, t) - \nu \frac{\partial^2}{\partial x^2} u(x, t) = -i \int dx' \chi(x - x') \mu(x't) ,$$
$$\frac{\partial}{\partial t} \mu(x, t) + u(x, t) \frac{\partial}{\partial x} \mu(x, t) + \nu \frac{\partial^2}{\partial x^2} \mu(x, t) = i\delta(t)\delta'(x) . \qquad (5.3.11)$$

The solution of these coupled system of partial differential equations is termed *instanton*. Accordingly, the instanton configuration corresponds to the most probable trajectory connecting initial conditions with the prescribed final condition. Hence, it can be interpreted as the classical trajectory of the path integral formulation of quantum mechanics in the limit $\hbar \to 0$. Therefore, the saddle point approximation in Eq. (5.3.8) is only justified in the presence of a small parameter. In particular, we should expect that the instanton approximation of the velocity gradient PDF in Eq. (5.3.8) becomes exact in the limit $a \to \infty$. In other words, instantons describe the *far-tail behavior* of PDFs. The latter characteristic of instantons makes them particularly suitable for a statistical description of turbulence, since they can be invoked to describe the extreme vorticity or velocity increment events.

The instanton method was first applied to turbulence by Gurarie and Migdal who determined the right tail of the velocity gradient PDF in Burgers turbulence [50].

Positive velocity gradients are associated to smooth ramps in between shocks and large positive gradients obey a gradient PDF

$$g(a) \sim \exp\left[\frac{2a^3}{3\chi''(0)}\right] \quad \text{for large positive } a, \qquad (5.3.12)$$

which is a result first obtained by Polyakov, who applied the operator product expansion to the PDF hierarchy in Burgers turbulence [51]. It has to be stressed that the above result is only valid for large a and makes no prediction of a possible algebraic pre-factor. Moreover, the second derivative of the spatial correlation of the forcing is negative, $\chi''(0) < 0$.

In comparison to the right tail of the velocity gradient PDF, the prediction for the left tail, i.e., the probability amplitude of shocks, is more complicated. Polyakov suggests a gradient PDF $g(a) \sim a^{-5/2}$ for large negative a that is once more based on the application of fusion rules [51]. By contrast, E and Vanden-Eijnden use a closure method for the dissipative term in the gradient PDF hierarchy that respects the geometrical aspects of shocks [52]. They conclude that the gradient PDF for negative a decays algebraically according to $g(a) \sim a^{-7/2}$. The instanton method devised by Balkovsky et al. [53] suggests

$$g(a) \sim \exp\left[-\frac{a^{3/2}}{\mathrm{Re}^{3/2}}\right] \quad \text{for large negative } a. \qquad (5.3.13)$$

Latter results seem to agree quite well with numerical evaluations of the instanton equations [54]. The application of the instanton method to three-dimensional turbulence is still an outstanding matter. For instance, a far-tail examination of the vorticity PDF via instantons would shed some more light on the peculiar vortical structures encountered in fully developed turbulence. Moreover, it would be of great interest to calculate fluctuations around the instanton solution (5.3.11) in order to access a broader range of the velocity gradient (or any other observable) than only the tails of the PDFs. Recently, a sampling scheme was proposed that explicitly used an evolution equation for fluctuations around instanton solutions in Burgers turbulence [55].

5.4 Dissipation Anomaly in Increment PDF Hierarchy of Burgers Turbulence

In this final section, we want to return to the discussion of the dissipation anomaly in turbulence which we already encountered in the context of shock solution of Burgers turbulence in Sect. 2.4 as well as in the section on multifractal predictions, Sect. 5.1. To this end, we again focus on the example of the Burgers equation 2.4.1

$$\frac{\partial}{\partial t}u(x,t) + u(x,t)\frac{\partial}{\partial x}u(x,t) = \nu\frac{\partial^2}{\partial x^2}u(x,t) + F(x,t), \tag{5.4.1}$$

with a white noise in time Gaussian forcing $F(x,t)$ defined by the second-order moment

$$\langle F(x,t)F(x',t)\rangle = \chi(x-x')\delta(t-t'), \tag{5.4.2}$$

where $\chi(x-x')$ is the spatial correlation function, assumed to be concentrated around a characteristic scale $|x-x'| \sim l_f$. The evolution equation for the velocity increment $v(x,r,t) = u(x+r,t) - u(x,t)$ is

$$\frac{\partial}{\partial t}v(x,r,t) + u(x,t)\frac{\partial v(x,r,t)}{\partial x} + v(x,r,t)\frac{\partial v(x,r,t)}{\partial r}$$
$$= \nu\frac{\partial^2 v(x,r,t)}{\partial x^2} + F(x+r,t) - F(x,t). \tag{5.4.3}$$

The temporal evolution of the one-increment PDF,

$$f_1(v_1,r_1,t) = \langle\delta(v_1 - v(x,r_1,t))\rangle, \tag{5.4.4}$$

is derived in Appendix 1 according to

$$\frac{\partial}{\partial t}f_1(v_1,r_1,t) + v_1\frac{\partial}{\partial r_1}f_1(v_1,r_1,t) + 2\int_{-\infty}^{v_1}dv'\frac{\partial}{\partial r_1}f_1(v',r_1,t)$$

$$= -\nu\frac{\partial}{\partial v_1}\int dr_2\,[\delta(r_2-r_1) - 2\delta(r_2)]\frac{\partial^2}{\partial r_2^2}\int dv_2 v_2 f_2(v_2,r_2;v_1,r_1,t)$$

$$+[\chi(0) - \chi(r_1)]\frac{\partial^2}{\partial v_1^2}f_1(v_1,r_1,t). \tag{5.4.5}$$

Due to the dissipative term, we are once more left with an infinite hierarchy of PDF equations similarly to the multi-point hierarchy discussed in Sect. 3.4. This hierarchy even prevails in the limit $\nu \to 0$, which is the dissipation anomaly in Burgers turbulence [51, 52]. In order to directly observe the implications of the dissipation anomaly, it is useful to reformulate dissipative terms: First, we assume the stationarity of velocity increment statistics, i.e., $\frac{\partial}{\partial t}f_1(v_1,r_1,t)=0$. Second, as shown in Appendix 2, the unclosed viscous term in Eq. (5.4.5) can be rewritten in terms of the joint velocity gradient and velocity increment statistics as

$$
\begin{aligned}
v_1 \frac{\partial}{\partial r_1} f_1(v_1, r_1) = &-2 \int_{-\infty}^{v_1} dv' \frac{\partial}{\partial r_1} f_1(v', r_1) + 2v \int_{-\infty}^{v_1} dv' \frac{\partial^2}{\partial r_1^2} f_1(v', r_1) \\
&- \frac{\partial^2}{\partial v_1^2} \left[\left\langle \frac{\varepsilon(x)}{2} [\delta(v_1 - v(x, r_1)) + \delta(v_1 + v(x, -r_1))] \right\rangle \right. \\
&\left. - [\chi(0) - \chi(r_1)] f_1(v_1, r_1) \right],
\end{aligned}
\tag{5.4.6}
$$

where we have introduced the local energy dissipation rate

$$
\varepsilon(x) = 2v \left(\frac{\partial u(x)}{\partial x} \right)^2.
\tag{5.4.7}
$$

By virtue of these transformations, we can see that the unclosed term in Eq. (5.4.5) involves joint statistics of the velocity increment and the velocity gradient. Taking the third-order moment of Eq. (5.4.6) yields

$$
\frac{1}{3} \frac{\partial}{\partial r_1} \langle v_1^3 \rangle = -2 \langle \varepsilon \rangle + 2v \frac{\partial^2}{\partial r_1^2} \langle v_1^2 \rangle + 2[\chi(0) - \chi(r_1)].
\tag{5.4.8}
$$

In the inertial range, we thus obtain

$$
\langle v_1^3 \rangle = -6 \langle \varepsilon \rangle r_1 \qquad \text{for} \quad \eta \ll r_1 \ll L,
\tag{5.4.9}
$$

where we have neglected forcing contributions. This can be considered as the Burgers 4/5-law from Sect. 3.3.3. Furthermore, we can specify the averaged local energy dissipation rate in terms of the forcing correlation function. Accordingly, we consider Eq. (5.4.8) for large r_1 in the range of the integral length scale. In this case, both the transport term on the l.h.s. and the viscous term vanish and we are left with $\langle \varepsilon \rangle = \chi(0)$, where we assumed that $\chi(r_1)$ tends to zero for large r_1. The rate of energy input on large scales $\chi(0)$ thus determines the averaged local rate of energy dissipation $\langle \varepsilon \rangle$. In the opposite case, i.e., for small r_1 the structure function of third order is negligible which yields

$$
-2 \langle \varepsilon \rangle + 2v \frac{\partial^2}{\partial r_1^2} \langle v_1^2 \rangle = 0.
\tag{5.4.10}
$$

Therefore, the second-order structure function in the dissipative range is given according to

$$\langle v_1^2 \rangle = \frac{\langle \varepsilon \rangle r_1^2}{2\nu} . \qquad (5.4.11)$$

5.4.1 Interpretation of the Dissipation Anomaly in the Moment Formulation

In the preceding section, we could relate the unclosed dissipative term in the evolution equation of the one-increment PDF (5.4.5) to a term that involves joint statistics of velocity increment and velocity gradient in Eq. (5.4.6). In this section, we want to interpret this dissipative term heuristically. To this end we take the moments $\langle v_1^n \rangle = \int dv_1^n f_1(v_1, r_1)$ of Eq. (5.4.6) which yields

$$\left(1 - \frac{2}{n}\right) \frac{\partial}{\partial r_1} \langle v(x, r_1)^n \rangle = 2\nu \frac{\partial^2}{\partial r_1^2} \langle v(x, r_1)^{n-1} \rangle$$
$$- \frac{(n-1)(n-2)}{2} \left\langle [v(x, r_1)^{n-3} + (-v(x, -r_1))^{n-3}] \varepsilon(x) \right\rangle$$
$$+ (n-1)(n-2)[\chi(0) - \chi(r_1)] \langle v(x, r_1)^{n-3} \rangle . \qquad (5.4.12)$$

In the following, we drop the forcing contribution as well as the (smooth) viscous contribution, which is reasonable in the inertial range. Furthermore, under the assumptions of a power law $\langle v_1^n \rangle \sim r_1^{\zeta_n}$ in the inertial range, we obtain

$$\left\langle v(x, r_1)^{n-3} \varepsilon(x) \right\rangle \sim r_1^{\zeta_n - 1} . \qquad (5.4.13)$$

Interestingly, this result is in accordance with predictions of the multifractal approach represented by Eq. (5.1.22). Hence, the dissipation anomaly in Burgers turbulence seems to agree with the multifractal approach that leads to the joint gradient-increment statistics in Eq. (5.1.22). The latter result should not come as surprise, since high Reynolds number Burgers turbulence (i.e., $\nu \to 0$ similar to the multifractal derivation in Sect. 5.1.2) is dominated by shock-like structures. At this point, we must stress that Eq. (5.4.13) does not further specify the power law behavior of structure functions $\langle v_1^n \rangle$. Therefore, it cannot be directly compared to phenomenological models of turbulence.

In the following, we seek to make certain heuristic assertions about the dissipative term in Eq. (5.4.12) via considering the underlying geometrical structures of the flow: In high Reynolds number Burgers turbulence, we are faced with shock-like structures similar to the one in Fig. 5.12a. In this case, the local energy dissipation rate peaks at the center of the shock and velocities $u(x + r_1)$ and $u(x - r_1)$ are arranged anti-symmetrically around $u(x) = 0$, which leads to Burgers scaling

(i.) Burgers scaling:

$$\langle [v(x, r_1)^{n-3} + (-v(x, -r_1))^{n-3}]\varepsilon(x) \rangle \sim u_{rms}^{n-3} \langle \varepsilon \rangle \sim r_1^{\zeta_n - 1}$$
$$\rightarrow \quad \zeta_n = 1 \,. \tag{5.4.14}$$

Let us now consider the Burgers equation with an additional nonlocality in the form of the Hilbert transform of the velocity field, which resembles the Navier-Stokes equation

$$\frac{\partial}{\partial t} u(x, t) + \alpha u(x, t) \frac{\partial}{\partial x} u(x, t) + \frac{1 - \alpha}{\pi} \text{p.v.} \int dx' \frac{u(x', t)}{x - x'} \frac{\partial}{\partial x} u(x, t) = \nu \frac{\partial^2}{\partial x^2} u(x, t) \,. \tag{5.4.15}$$

This equation exhibits intermittency behavior that ranges from self-similar behavior to shock-like behavior depending on a balance between nonlinearity and nonlocality adjusted by the parameter α [56]. Here, dominant structures are so-called cusps similar to the one depicted in Fig. 5.12b. For this case, the velocity field possesses the symmetry $u(x - r_1) = -u(x + r_1)$ leading to the disappearance of the dissipative term for even n. Accordingly, the power law behavior of the structure functions will be determined by the balance of the nonlinearity and the forcing term as predicted by the renormalization group from Sect. 4.5.2. Another important consequence of Eq. (5.4.12) is the case when local dissipation rate and the velocity increment are statistically independent.

(ii.) Kolmogorov's mean-field theory:

$$\langle [v(x, r_1)^{n-3} + (-v(x, -r_1))^{n-3}]\varepsilon(x) \rangle$$
$$= \underbrace{\langle [v(x, r_1)^{n-3} + (-v(x, -r_1))^{n-3}] \rangle}_{\sim r_1^{\zeta_{n-3}}} \langle \varepsilon(x) \rangle \sim r_1^{\zeta_n - 1}$$
$$\rightarrow \quad \zeta_n - 1 = \zeta_{n-3} \quad \rightarrow \quad \zeta_n = n/3 \,. \tag{5.4.16}$$

Hence, statistical independence of the energy dissipation rate and the velocity increment necessarily implies K41 scaling. This conjecture is in accordance with Kolmogorov's second similarity hypothesis and underlines the importance of Landau's objection (see [6]): Landau recognized that Kolmogorov's theory solely depends on the *averaged* local energy dissipation rate and completely neglects a possible random character of the energy transfer. Accordingly, deviations from K41 scaling can be attributed to non-vanishing statistical correlations between the energy dissipation

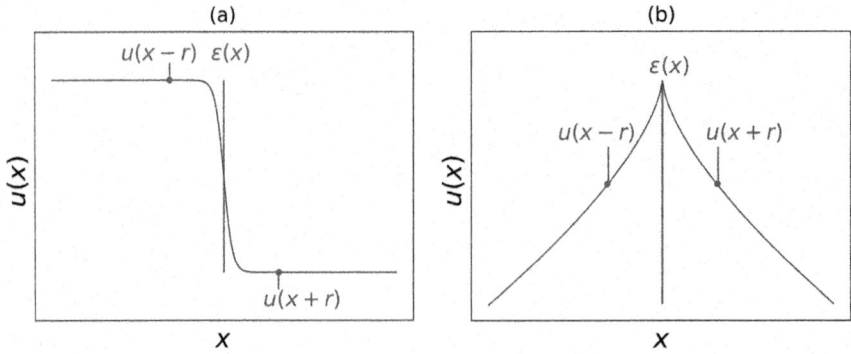

Fig. 5.12 a Schematic depiction of a shock in Burgers turbulence. The local energy dissipation rate is peaked at the center of the shock $\varepsilon(x)$. Depending on the strength of the shock, the velocity field at $u(x - r)$ and at $u(x + r)$ possesses the symmetry $u(x - r) = -u(x + r)$, which leads to Burgers scaling (5.4.14). **b** In the case of cusp-like structures, $\varepsilon(x)$ is still peaked in the center of the cusp. The symmetry of the cusps, however, leads to the vanishing of the dissipation anomaly in Eq. (5.4.12)

rate (or the velocity gradient) and velocity increments at different scales. Henceforth, Burgers scaling (5.4.14) must be considered as the complete opposite case: here, the energy dissipation rate fully correlates with the velocity increment, which results in pronounced intermittency effects.

5.4.2   Gurarie-Migdal Solution for the Right Tail of the One-Increment PDF

In this section, we consider Eq. (5.4.5) for positive velocity increments. For this case, it is appropriate to neglect the influence of the viscous term, since we have to deal primarily with smooth ramps in between shocks. Accordingly, we obtain a closed equation for the one-increment PDF

$$v_1 \frac{\partial}{\partial r_1} f_1(v_1, r_1) = -2 \int_{-\infty}^{v_1} dv' \frac{\partial}{\partial r_1} f_1(v', r_1) + [\chi(0) - \chi(r_1)] \frac{\partial^2}{\partial v_1^2} f_1(v_1, r_1) .$$

(5.4.17)

We approximate $\chi(0) - \chi(r_1) \approx -\frac{1}{2}\chi''(0)r_1^2$, which corresponds to a Burgers equation with forcing linear in x [57]. Furthermore, a self-similar ansatz for the PDF,

$$f_1(v_1, r_1) = \frac{1}{r_1} g\left(\frac{v_1}{r_1}\right) ,$$

(5.4.18)

yields

$$\frac{\partial}{\partial r_1} f_1(v_1, r_1) = -\frac{1}{r_1^2}[g(\xi_1) + \xi_1 g'(\xi_1)] \tag{5.4.19}$$

$$\frac{\partial^2}{\partial v_1^2} f_1(v_1, r_1) = \frac{1}{r_1^3} g''(\xi_1) \tag{5.4.20}$$

$$\int_{-\infty}^{v_1} dv \frac{\partial}{\partial r_1} f_1(v, r_1) = -\frac{1}{r_1} \xi_1 g(\xi_1) , \tag{5.4.21}$$

where $\xi_1 = v_1/r_1$. Inserting these quantities into (5.4.17) yields an ordinary differential equation for $g(\xi_1)$, namely,

$$-\frac{\chi''(0)}{2} g''(\xi_1) + \xi_1^2 g'(\xi_1) + 3\xi_1 g(\xi_1) = 0 . \tag{5.4.22}$$

The solution of this differential equation is

$$g(\xi_1) = C_1 \xi_1 \exp\left[\frac{2\xi_1^3}{3\chi''(0)}\right]$$

$$+ C_2 \exp\left[\frac{2\xi_1^3}{3\chi''(0)}\right] \left(-\frac{\exp\left[\frac{2\xi_1^3}{3\chi''(0)}\right]}{\xi_1} + \frac{\left(\frac{2\xi_1^3}{3\chi''(0)}\right)^{\frac{1}{3}} \Gamma\left(\frac{2}{3}, \frac{2\xi_1^3}{3\chi''(0)}\right)}{\xi_1}\right) , \tag{5.4.23}$$

where Γ denotes the incomplete gamma function. It is far from obvious how the constants C_1 and C_2 are fixed, since we only considered positive increments and neglected the dissipative term. However, for the special case $C_1 = 0$, we get

$$f_1(v_1, r_1) = C_1 \frac{v_1}{r_1} \exp\left[\frac{2v_1^3}{3r_1^3 \chi''(0)}\right] \quad \text{for positive } v_1 . \tag{5.4.24}$$

For large v_1, i.e., the right tail of the one-increment PDF, we recover the Gurarie-Migdal solution (5.3.12) discussed in Sect. 5.3. Here, we obtained an algebraic correction v_1/r_1 to the instanton solution.

5.5 Chapter Conclusions

This chapter highlighted the importance of non-perturbative methods for the realm of a statistical description for the Navier-Stokes equation. Starting with the operator product expansion in Sect. 5.1, it was shown how multi-point statistics could be simplified by the assumption of uncorrelated multipliers of the underlying stochastic

processes. A similar treatment has been discussed in Sect. 5.2 with the help of a quite
intuitive interpretation of the turbulent energy cascade as a Markov process of velocity
increments in scale. In fact, we demonstrated that the fusion rules in Sect. 5.1 are a
direct consequence of the Markov property. The inaccuracy of the fusion rules for
small relative scale separations can therefore be interpreted as the breakdown of the
Markov property at the Markov-Einstein length. Furthermore, the phenomenology
by Friedrich and Peinke [43] is capable of capturing the essence of anomalous scaling
embodied in Eq. (5.2.33). An admissible description of intermittency in turbulence,
however, should take into account an infinite number of Kramers-Moyal coefficients,
which has been demonstrated by the semi-logarithmic plot of the reduced Kramers-
Moyal coefficients in Fig. 5.3b. This opens up the possibility to decide which of the
various phenomenological models is best suited to describe a given turbulent flow.

Moreover, since this is a purely phenomenological description of turbulence, it
is appropriate to invoke the basic fluid dynamical equations. The Markov property
can thus be considered as an effective three-point closure of the infinite hierarchy
of the multi-point PDF hierarchy discussed in Sect. 3.4. The hope in this ongoing
investigation [40] is that, in contrast to the closure methods discussed in Chap. 4, the
Markov property will preserve the intermittent character of the velocity field. Furthermore, it is of great importance to combine the latter approach with the instanton
method described in Sect. 5.3. In this scenario, the Markov property might be used
for closing the hierarchy at a given order and deploy the instanton method in order
to approximate a higher order moment contained in the dissipation anomaly.

Appendix 1: Derivation of Multi-increment Hierarchy in Burgers Turbulence

In order to derive the evolution equation (5.4.5) we take the temporal derivative of
the one-increment PDF

$$
\begin{aligned}
\frac{\partial}{\partial t} f_1(v_1, r_1, t) &= \frac{\partial}{\partial t} \langle \delta(v_1 - v(x, r_1, t)) \rangle \\
&= -\frac{\partial}{\partial v_1} \left\langle \delta(v_1 - v(x, r_1, t)) \frac{\partial}{\partial t} v(x, r_1, t) \right\rangle \\
&= \frac{\partial}{\partial v_1} \left\langle \delta(v_1 - v(x, r_1, t)) \left[u(x, t) \frac{\partial}{\partial x} v(x, r_1, t) + v(x, r_1, t) \frac{\partial}{\partial r_1} v(x, r_1, t) \right. \right. \\
&\quad \left. \left. -\nu \frac{\partial^2}{\partial x^2} v(x, r_1, t) - F(x + r_1, t) + F(x, t) \right] \right\rangle,
\end{aligned}
\tag{5.5.1}
$$

where Eq. (5.4.3) was used to replace the temporal evolution of the velocity increment. Each term can now be treated separately. Starting with the first advective term,
we obtain

$$\frac{\partial}{\partial v_1}\left\langle \delta(v_1 - v(x, r_1, t))v(x, t)\frac{\partial}{\partial x}v(x, r_1, t)\right\rangle = \left\langle v(x, t)\frac{\partial}{\partial x}\delta(v_1 - v(x, r_1, t))\right\rangle$$

$$= \underbrace{\frac{\partial}{\partial x}\langle u(x, t)\delta(v_1 - v(x, r_1, t))\rangle}_{=0,\ \text{homogeneity}} - \underbrace{\left\langle \frac{\partial u(x, t)}{\partial x}\delta(v_1 - v(x, r_1, t))\right\rangle}_{=\left[\frac{\partial v(x,r_1,t)}{\partial r_1} - \frac{\partial v(x,r_1,t)}{\partial x}\right]\times\delta}$$

$$= \int_{-\infty}^{v_1} dv_1' \frac{\partial}{\partial r_1}\underbrace{\langle \delta(v_1' - v(x, r_1, t))\rangle}_{=f_1(v_1',r,t)} - \int_{-\infty}^{v_1} dv_1' \frac{\partial}{\partial x}\underbrace{\langle \delta(v_1' - v(x, r_1, t))\rangle}_{=0,\ \text{homogeneity}} \quad . \quad (5.5.2)$$

In the first and last steps, we made use of the inverse chain rule. The second advective term can be treated in the same way according to

$$-\frac{\partial}{\partial v_1}\left\langle \delta(v_1 - v(x, r_1, t))v(x, r_1, t)\frac{\partial}{\partial r_1}v(x, r_1, t)\right\rangle$$

$$= \left\langle v(x, r_1, t)\frac{\partial}{\partial r_1}\delta(v_1 - v(x, r_1, t))\right\rangle$$

$$= \frac{\partial}{\partial r_1}\langle v(x, r_1, t)\delta(v_1 - v(x, r_1, t))\rangle - \left\langle \frac{\partial v(x, r_1, t)}{\partial r_1}\delta(v_1 - v(x, r_1, t))\right\rangle$$

$$= v_1\frac{\partial}{\partial r_1}\underbrace{\langle \delta(v_1 - v(x, r_1, t))\rangle}_{=f_1(v_1,r_1,t)} + \int_{-\infty}^{v_1} dv_1' \frac{\partial}{\partial r_1}\underbrace{\langle \delta(v_1' - v(x, r_1, t))\rangle}_{=f_1(v_1',r_1,t)}, \quad (5.5.3)$$

where we applied the sifting property of the δ-function, i.e., $v(x, r_1, t)\delta(v_1 - v(x, r_1, t)) = v_1\delta(v_1 - v(x, r_1, t))$. Nonlinear terms can thus be expressed solely in terms of the one-increment PDF or its associated cumulative PDF, which is a particularity of the Burgers equation (for the Navier-Stokes equation we would be facing unclosed terms from the pressure [58]). However, viscous contributions in Eq. (5.5.1) confront us with unclosed terms and we have to introduce the two-increment PDF, which results in an infinite hierarchy of PDF equations. This can be seen from the calculation of the viscous term in Eq. (5.5.1),

$$-\nu\left\langle \delta(v_1 - v(x, r_1, t))\frac{\partial^2 v(x, r_1, t)}{\partial x^2}\right\rangle$$

$$-\ -\nu\left\langle \delta(v_1 - v(x, r_1, t))\left[\frac{\partial^2 v(x, r_1, t)}{\partial r_1^2} - \frac{\partial^2 v(x, t)}{\partial x^2}\right]\right\rangle$$

$$= -\nu\int dr_2\,[\delta(r_2 - r_1) - \delta(r_2)]\frac{\partial^2}{\partial r_2^2}\langle v(x, r_2, t)\delta(v_1 - v(x, r_1, t))\rangle$$

$$= -\nu \int dr_2 \left[\delta(r_2 - r_1) - \delta(r_2)\right] \frac{\partial^2}{\partial r_2^2}$$

$$\times \int dv_2 v_2 \underbrace{\langle \delta(v_2 - v(x, r_2, t))\delta(v_1 - v(x, r_1, t))\rangle}_{=f_2(v_2, r_2; v_1, r_1, t)}. \tag{5.5.4}$$

We can handle forcing contributions in Eq. (5.5.1) by the usual trick of the Langevin equation [18]. Inserting the above calculations yields the evolution equation for the one-increment PDF

$$\frac{\partial}{\partial t} f_1(v_1, r_1, t) + v_1 \frac{\partial}{\partial r_1} f_1(v_1, r_1, t) + 2 \int_{-\infty}^{v_1} dv_1' \frac{\partial}{\partial r_1} f_1(v_1', r_1, t)$$

$$= -\nu \frac{\partial}{\partial v_1} \int dr_2 \left[\delta(r_2 - r_1) - \delta(r_2)\right] \frac{\partial^2}{\partial r_2^2} \int dv_2 v_2 f_2(v_2, r_2; v_1, r_1, t)$$

$$+ [\chi(0) - \chi(r_1)] \frac{\partial^2}{\partial v_1^2} f_1(v, r_1, t). \tag{5.5.5}$$

Appendix 2: Reformulation of the Viscous Term in the Multi-increment Hierarchy of Burgers Turbulence

In this appendix, we show that the unclosed term in the evolution equation of the one-increment PDF (5.4.5) involves the local energy dissipation rate. To this end, we rewrite viscous contributions in their original form according to

$$\nu \int dr_2 \left[\delta(r_2 - r_1) - \delta(r_2)\right] \frac{\partial^2}{\partial r_2^2} \int dv_2 v_2 f_2(v_2, r_2; v_1, r_1)$$

$$= \nu \left\langle \left[\frac{\partial^2 v(x, r_1)}{\partial r_1^2} - \frac{\partial^2 u(x)}{\partial x^2}\right] \delta(v_1 - v(x, r_1))\right\rangle. \tag{5.5.6}$$

A further treatment of these terms yields

$$+\nu \frac{\partial}{\partial r_1}\left\langle \frac{\partial v(x, r_1)}{\partial r_1}\delta(v_1 - v(x, r_1))\right\rangle - \nu\left\langle \frac{\partial v(x, r_1)}{\partial r_1}\frac{\partial\delta(v_1 - v(x, r_1))}{\partial r_1}\right\rangle$$

$$\underbrace{-\nu \frac{\partial}{\partial x}\left\langle \frac{\partial u(x)}{\partial x}\delta(v_1 - v(x, r_1))\right\rangle}_{=0,\ \text{homogeneity}} +\nu\left\langle \frac{\partial u(x)}{\partial x}\frac{\partial\delta(v_1 - v(x, r_1))}{\partial x}\right\rangle$$

$$= -\nu \int_{-\infty}^{v_1} dv_1' \frac{\partial^2}{\partial r_1^2}\langle\delta(v_1' - v(x, r_1))\rangle + \nu\frac{\partial}{\partial v_1}\left\langle \left(\frac{\partial v(x, r_1)}{\partial r_1}\right)^2 \delta(v_1 - v(x, r_1))\right\rangle$$

$$-\nu\frac{\partial}{\partial v_1}\left\langle \underbrace{\frac{\partial u(x)}{\partial x}\left(\frac{\partial v(x, r_1)}{\partial x}\right)}_{=\frac{\partial v(x,r_1)}{\partial r_1}-\frac{\partial u(x)}{\partial x}} \delta(v_1 - v(x, r_1))\right\rangle. \tag{5.5.7}$$

Inserting the one-increment PDF $f_1(v_1', r_1)$ into the first term on the r.h.s. yields

$$
-\nu \int_{-\infty}^{v_1} dv_1' \frac{\partial^2}{\partial r_1^2} f_1(v_1', r_1) + \nu \frac{\partial}{\partial v_1} \left\langle \left(\frac{\partial v(x+r_1)}{\partial r_1} \right)^2 \delta(v_1 - v(x, r_1)) \right\rangle
$$

$$
+\nu \frac{\partial}{\partial x} \underbrace{\left\langle \left(\frac{\partial v(x)}{\partial x} \right) \delta(v_1 - v(x, r_1)) \right\rangle + \nu \frac{\partial}{\partial v_1} \left\langle \left(\frac{\partial v(x)}{\partial x} \right)^2 \delta(v_1 - v(x, r_1)) \right\rangle}_{=\left(\frac{\partial v(x,r_1)}{\partial r_1} - \frac{\partial v(x,r_1)}{\partial x} \right) \times \delta}
$$

$$
= -\nu \int_{-\infty}^{v_1} dv_1' \frac{\partial^2}{\partial r_1^2} f_1(v_1', r_1) + \nu \frac{\partial}{\partial v_1} \left\langle \left(\frac{\partial v(x+r_1)}{\partial r_1} \right)^2 \delta(v_1 - v(x, r_1)) \right\rangle
$$

$$
-\nu \int_{-\infty}^{v_1} dv_1' \frac{\partial^2}{\partial r_1^2} f_1(v_1', \dot{r}) + \nu \int_{-\infty}^{v_1} dv_1' \underbrace{\frac{\partial^2}{\partial r_1 \partial x} f_1(v_1', r_1)}_{=0, \text{ homogeneity}}
$$

$$
+\nu \frac{\partial}{\partial v_1} \left\langle \left(\frac{\partial v(x)}{\partial x} \right)^2 \delta(v_1 - v(x, r_1)) \right\rangle . \tag{5.5.8}
$$

Under the assumption of homogeneity, we obtain

$$
\left\langle \left(\frac{\partial v(x+r_1)}{\partial r_1} \right)^2 \delta(v_1 - v(x, r_1)) \right\rangle
$$

$$
= \left\langle \left(\frac{\partial v(x+r_1)}{\partial r_1} \right)^2 \delta(v_1 - v(x + r_1) + v(x)) \right\rangle
$$

$$
= \left\langle \left(\frac{\partial v(x)}{\partial x} \right)^2 \delta(v_1 - u(x) + u(x - r_1)) \right\rangle = \left\langle \left(\frac{\partial v(x)}{\partial x} \right)^2 \delta(v_1 + v(x, -r_1)) \right\rangle, \tag{5.5.9}
$$

which allows the introduction of the local energy dissipation rate in Eq. (5.4.5) according to

$$
v_1 \frac{\partial}{\partial r_1} f_1(v_1, r_1) = -2 \int_{-\infty}^{v_1} dv' \frac{\partial}{\partial r_1} f_1(v', r_1) + 2\nu \int_{-\infty}^{v_1} dv' \frac{\partial^2}{\partial r_1^2} f_1(v', r_1)
$$

$$
- \frac{\partial^2}{\partial v_1^2} \left[\left\langle \frac{\varepsilon(x)}{2} [\delta(v_1 - v(x, r_1)) + \delta(v_1 + v(x, -r_1))] \right\rangle
$$

$$
- [\chi(0) - \chi(r_1)] f_1(v_1, r_1) \right] . \tag{5.5.10}
$$

References

1. Bonneau, G.: Operator product expansion. Scholarpedia **4**(9), 8506 (2009)
2. Weinberg, S.: Current algebra and gauge theories. i. Phys. Rev. D **8**(2), 605 (1973)
3. Wilson, K.G.: Non-lagrangian models of current algebra. Phys. Rev. **179**(5), 1499–1512 (1969)
4. Paladin, G., Vulpiani, A.: Degrees of freedom of turbulence. Phys. Rev. A **35**(4), 1971–1973 (1987)
5. Landau, L.D., Lifshitz, E.M.: Physics, Third Edition: Volume 5 (Course of Theoretical Physics). Butterworth-Heinemann (1987)
6. Frisch, U.: Turbulence. Cambridge University Press (1995)
7. Frisch, U., Vergassola, M.: A prediction of the multifractal model: the intermediate dissipation range. Europhys. Lett. **14**(5), 439 (1991)
8. Nelkin, M.: Multifractal scaling of velocity derivatives in turbulence. Phys. Rev. A **42**(12), 7226–7229 (1990)
9. Anselmet, F., Gagne, Y., Hopfinger, E.J., Antonia, R.A.: High-order velocity structure functions in turbulent shear flows. J. Fluid Mech. **140**, 63–89 (1984)
10. Kolmogorov, A.N.: A refinement of previous hypotheses concerning the local structure of turbulence in a viscous incompressible fluid at high Reynolds number. J. Fluid Mech. **13**(01), 82–85 (1962)
11. Oboukhov, A.M.: Some specific features of atmospheric tubulence. J. Fluid Mech. **67**(8), 77–81 (1962)
12. Benzi, R., Biferale, L., Toschi, F.: Multiscale velocity correlations in turbulence. Phys. Rev. Lett. **80**(15), 3244–3247 (1998)
13. Benzi, R., Biferale, L., Ruiz-Chavarria, G., Ciliberto, S., Toschi, F.: Multiscale velocity correlation in turbulence: Experiments, numerical simulations, synthetic signals. Phys. Fluids **11**(8), (1999)
14. Eyink, G.L.: Lagrangian field theory, multifractals, and universal scaling in turbulence. Phys. Lett. A **172**(5), 355–360 (1993)
15. L'vov, V. and Procaccia, I.: Fusion rules in turbulent systems with flux equilibrium. Phys. Rev. Lett. **76**(16), 2898–2901 (1996)
16. Friedrich, R., Peinke, J.: Description of a turbulent cascade by a fokker-planck equation. Phys. Rev. Lett. **78**(5), 863–866 (1997)
17. Friedrich, R., Peinke, J., Tabar, R.M.: Importance of fluctuations: complexity in the view of stochastic processes. Encycl. Complex. Syst. Sci. **21**(1982):Entry 294 (2009)
18. Risken, H.: The Fokker-Planck Equation. Springer, Berlin (1996)
19. Renner, C., Peinke, J., Friedrich, R.: Experimental indications for Markov properties of small-scale turbulence. J. Fluid Mech. **433**, 383–409 (2001)
20. Renner, C.: Markowanalysen stochastisch fluktuierender Zeitserien. PhD thesis, Carl von Ossietzky Universität Oldenburg (2002)
21. Voßkuhle, M.: Statistische Analysen zweidimensionaler Turbulenz. PhD thesis, University of Münster (2009)
22. Lück, S., Renner, C., Peinke, J., and Friedrich, R.: The Markov-Einstein coherence length-a new meaning for the Taylor length in turbulence. Phys. Lett. Sect. A Gen. At. Solid State Phys. 359(5):335–338 (2006)
23. Einstein, A.: On the movement of small particles suspended in stationary liquids required by the molecular-kinetic theory of heat. Ann. Phys. **17**, 549–560 (1905)
24. Srinivas, M.D., Wolf, E.: Stochastic Equations for Classical and Quantum Distribution Functions. Springer, US (1977)
25. Gardiner, C.W.: Handbook of Stochastic Methods. Springer, Berlin (1983)
26. Friedrich, J., Margazoglou, G., Biferale, L., Grauer, R.: Multiscale velocity correlations in turbulence and Burgers turbulence: Fusion rules, Markov processes in scale, and multifractal predictions. Phys. Rev. E **98**(2), 023104 (2018)
27. Kolmogorov, A.N.: The local structure of turbulence in incompressible viscous fluid for very large Reynolds numbers. Dokl. Akad. Nauk SSSR **30**(1890), 301–305 (1941)

28. Bec, J., Khanin, K.: Burgers turbulence. Phys. Rep. 447(1–2):1–66 (2007)
29. She, Z.-S., Leveque, E.: Universal scaling laws in fully developed turbulence. Phys. Rev. Lett. **72**(3), 336–339 (1994)
30. Nickelsen, D.: Master equation for she-leveque scaling and its classification in terms of other markov models of developed turbulence. J. Stat. Mech: Theory Exp. **2017**(7), 073209 (2017)
31. Yakhot, V.: Probability density and scaling exponents of the moments of longitudinal velocity difference in strong turbulence. Phys. Rev. E **57**(2), 1737–1751 (1998)
32. Yakhot, V.: Mean-field approximation and a small parameter in turbulence theory. Phys. Rev. E **63**, 26307 (2001)
33. Yakhot, V.: Probability densities in strong turbulence. Phys. D **215**, 166–174 (2006)
34. Novikov, E.A.: Infinitely divisible distributions in turbulence. Phys. Rev. E **50**(5), R3303–R3305 (1994)
35. Castaing, B.: The temperature of turbulent flows. J. Phys. II Fr. **6**(1), 105–114 (1996)
36. Eling, C., Oz, Y.: The anomalous scaling exponents of turbulence in general dimension from random geometry. J. High Energy Phys. **2015**(9), (2015)
37. Pawula, R.F.: Approximation of the linear boltzmann equation by the fokker-planck equation. Phys. Rev. **162**(1), 186–188 (1967)
38. Courant, R., Hilbert, D.: Methods of Mathematical Physics II. Wiley (1962)
39. Castaing, B., Gagne, Y., Hopfinger, E.J.: Velocity probability density functions of high Reynolds number turbulence. Phys. D Nonlinear Phenom. **46**(2), 177–200 (1990)
40. Friedrich, J.: Closure of the Lundgren-Monin-Novikov hierarchy in turbulence via a Markov property of velocity increments in scale. PhD thesis, Ruhr-University Bochum (2017)
41. Homann, H.: Lagrange Statistics of turbulent Flows in Fluids and Plasmas. Phd thesis, Ruhr-Universität Bochum (2006)
42. Grauer, R., Homann, H., Pinton, J.-F.: Longitudinal and transverse structure functions in high-Reynolds-number turbulence. New J. Phys. **14**, 63016 (2012)
43. Friedrich, R., Zeller, J., Peinke, J.: A note on three-point statistics of velocity increments in turbulence. EPL (Eur. Lett.) **41**(2), 153 (1998)
44. Hellinger, E.: Neue Begründung der Theorie quadratischer Formen von unendlichvielen Veränderlichen. J. fur die Reine und Angew. Math. **1909**(136), 210–271 (1909)
45. Friedrich, R., Renner, C., Siefert, M., Peinke, J.: Comment on "Indispensable Finite Time Corrections for Fokker-Planck Equations from Time Series Data". Phys. Rev. Lett. **89**(14), 149401 (2002)
46. Grafke, T., Grauer, R., Schäfer, T.: The instanton method and its numerical implementation in fluid mechanics. J. Phys. A: Math. Theor. **48**(33), 333001 (2015)
47. Martin, P.C., Siggia, E.D., Rose, H.A.: Statistical dynamics of classical systems. Phys. Rev. A **8**(1), 423–437 (1973)
48. Ivashkevich, E.V.: Symmetries of the stochastic Burgers equation. J. Phys. A: Math. Gen. **30**(15), L525 (1997)
49. Onsager, L., Machlup, S.: Fluctuations and Irreversible Processes. Phys. Rev. **91**(6), 1505–1512 (1953)
50. Gurarie, V., Migdal, A.: Instantons in the burgers equation. Phys. Rev. E **54**(5), 4908–4914 (1996)
51. Polyakov, A.M.: Turbulence without pressure. Phys. Rev. E **52**(6), 6183–6188 (1995)
52. E, W., Vanden Eijnden, E.,: Asymptotic theory for the probability density functions in burgers turbulence. Phys. Rev. Lett. **83**(13), 2572–2575 (1999)
53. Balkovsky, E., Falkovich, G., Kolokolov, I., Lebedev, V.: Intermittency of Burgers' Turbulence. Phys. Rev. Lett. **78**(8), 1452–1455 (1997)
54. Chernykh, A.I., Stepanov, M.G.: Large negative velocity gradients in Burgers turbulence. Phys. Rev. E **64**(2), 26306 (2001)
55. Ebener, L., Margazoglou, G., Friedrich, J., Biferale, L., Grauer, R.: Instanton based importance sampling for rare events in stochastic PDEs. Chaos **29**(6), 063102 (2019)
56. Zikanov, O., Thess, A., Grauer, R.: Statistics of turbulence in a generalized random-force-driven Burgers equation. Phys. Fluids **9**(5), 1362 (1997)

57. Eule, S., Friedrich, R. A note on the forced Burgers equation. Phys. Lett. Sect. A Gen. At. Solid State Phys. **351**(4-5):238–241 (2006)
58. Ulinich, F.R., Lyubimov, B.Y.: The statistical theory of turbulence of an incompressible fluid at large Reynolds numbers. Sov. J. Exp. Theor. Phys. **28**, 494 (1969)

Chapter 6
Outlook

The longstanding problem of turbulence still awaits a thorough and complete understanding of the equations that govern turbulent fluid motion. As it has been stressed throughout this monograph, the overwhelmingly complex spatio-temporal organization of turbulent flows requires a comprehensive statistical treatment of the Navier-Stokes equation. Latter spatio-temporal organization, which also becomes apparent from the existence of scaling laws, can be considered as a key signature of a strongly interacting system with vast numbers of degrees of freedom. The empirical evidence for small-scale intermittency, i.e., fluctuations that possess non-Gaussian properties at small scales, requires further investigation and descriptions that are not solely based on phenomenologies. It would therefore be desirable to infer certain characteristics directly from the underlying equations. The hierarchical character of the corresponding multi-point statistics of the Navier-Stokes equation, however, still has to be dealt with in an adequate fashion. Perturbative closure methods that are based on Gaussian or quasi-Gaussian assumptions have to be considered as mere formal treatments since perturbation expansion is set up in powers of the Reynolds numbers. The lack of a small parameter in turbulence theory, which would allow for the elimination of fast varying variables from the equations, further proves that a turbulent flow possesses no clearly distinguishable spatio-temporal scales. Further related questions are as follows:

- What is the impact of empirically observed elongated vortical structures on the energy transport in a turbulent flow?
- What is the relation between such coherent structures and statistical descriptions of turbulence by means of multi-point statistical quantities?
- How does an appropriate treatment of unclosed terms in these multi-point hierarchies look like? What assumptions have to be introduced in order to arrive at a closed system of equations (at a given order) and how can the latter be solved adequately?

© Springer Nature Switzerland AG 2021   161
J. Friedrich, *Non-perturbative Methods in Statistical Descriptions of Turbulence*,
Progress in Turbulence - Fundamentals and Applications 1,
https://doi.org/10.1007/978-3-030-51977-3_6

These open questions suggest that a comprehensive framework of turbulence theory does not yet exist. Therefore, non-perturbative methods that were discussed in Chap. 5 have to be further refined and advanced. The instanton method, for instance, has to be tested for the case of three-dimensional turbulence, thereby calculating extreme events of vorticity and so forth. Moreover, fluctuations around the instanton solution might shed further light on the far-tail behavior of velocity gradients which has to be tested on the basis of Burgers equation. First promising approaches that address these outstanding issues already exist. Furthermore, as it was shown in Sec. 5.2.3, seemingly different concepts such as the operator product expansion and the Markov approach might be reconciled in order to give a more coherent picture of the problem and to make further progress. Therefore, it will be exciting to witness the next decades of turbulence research where a combination of increased high Reynolds number numerical simulations, advanced subgrid-scale models for large eddy simulations, as well as further improved stochastic models will certainly advance our understanding of turbulent flows considerably.

Index

© Springer Nature Switzerland AG 2021
J. Friedrich, *Non-perturbative Methods in Statistical Descriptions of Turbulence*,
Progress in Turbulence - Fundamentals and Applications 1,
https://doi.org/10.1007/978-3-030-51977-3

Printed by Printforce, United Kingdom